*The San Marcos*

RIVER BOOKS

*Sponsored by the*

 River Systems Institute
at Texas State University

Andrew Sansom, General Editor

A&M nature guides

# THE SAN MARCOS
## A River's Story

BY JIM KIMMEL

Photographs by Jerry Touchstone Kimmel

TEXAS A&M UNIVERSITY PRESS   COLLEGE STATION

Copyright © 2006
by Jim Kimmel
Printed in China by Everbest Printing Co.
through Four Colour Imports
All rights reserved
First edition

The paper used in this book
meets the minimum requirements
of the American National Standard
for Permanence of Paper
for Printed Library Materials,
z39.48-1984.
Binding materials have been
chosen for durability.

LIBRARY OF CONGRESS
CATALOGING-IN-PUBLICATION DATA

Kimmel, Jim, 1943–
  The San Marcos : a river's story / by Jim Kimmel ;
photogaphs by Jerry Touchstone Kimmel.—1st ed.
     p.  cm.
  Includes bibliographical references and index.
  ISBN-13: 978-1-58544-542-4 (cloth : alk. paper)
  ISBN-10: 1-58544-542-8 (cloth : alk. paper)
   1.  San Marcos River (Tex.)—Description
and travel.   2.  San Marcos River Region
(Tex.)—History, Local.   I.  Title.
F392.S236K56   2006
976.4'31—dc22        2006001540

*This book
is dedicated to
Andrew Sansom
& the people of the
San Marcos River.*

# CONTENTS

|  | Foreword, by Andrew Sansom | IX |
|---|---|---|
|  | Preface | XI |
| Chapter 1 | *More than Just a Little River* | 1 |
| Chapter 2 | *Life of the River* | 15 |
| Chapter 3 | *From the Deep Past at the Springs* | 31 |
| Chapter 4 | *Anglo Americans at the Springs* | 49 |
| Chapter 5 | *San Marcos the River, San Marcos the Town* | 67 |
| Chapter 6 | *Dams and Towns: A Nostalgic River Landscape* | 91 |
| Chapter 7 | *Past Creates Future* | 117 |
| Appendix 1 | *Native Species in the Upper San Marcos River* | 131 |
| Appendix 2 | *Nonnative Species in the Upper San Marcos River* | 135 |
| Appendix 3 | *Summary of Edwards Aquifer Habitat Conservation Plan Measures* | 139 |
|  | References | 141 |
|  | Index | 149 |

# FOREWORD

The San Marcos is an iconic river in Texas. Maybe it is because this river has been known to people longer than any other, its headwaters thought to be the site of some of the oldest human settlements on the continent. Maybe it is because the river has maintained its spectacular natural beauty while contributing to the economic development of its basin and beyond, providing power for early industry and water for farms and cities. Most likely, it is because hundreds of thousands of Texans have experienced the river's wonders firsthand in the past century as visitors to Aquarena Springs at the river's source.

Jim Kimmel tells the story of the San Marcos by creating a rich blend of the river's natural and cultural history. In a gentle but learned way, one of Texas' most distinguished geographers helps us to understand and appreciate its great beauty and diversity and also connect with the people who have lived on its banks. Jerry Touchstone Kimmel's sensational photography vivifies the account, and thus these pages yield a full and satisfying picture of the character of this lovely waterway.

Today, the San Marcos story is emblematic of the larger struggle to protect all of Texas' rivers. The probability of increased pumping from the Edwards Aquifer threatens spring flows, which sustain the river, and risks continued impairments to the recharge of the aquifer. Currently, there are no laws to adequately ensure that once all of the needs and demands for water from the San Marcos are met, there will still be water flowing down the stream to give life to its unique ecosystems and feeding into San Antonio Bay, which depends on it for nourishment.

Thanks to Jim Kimmel, we can here find inspiration from Edward Burleson, who built the first dam; A. B. Rogers, who turned the resulting spectacular lake into one of the most successful tourism ventures in Texas history; or Jerome Supple, former president of the great university that thrives alongside the river. In the face of formidable obstacles and strong opposition, Supple engineered the purchase of the San Marcos River headwaters for their permanent protection and for the rich educational and recreational opportunities they provide.

Thus, the Texas State University family continues to be inspired by the river, dedicated to its stewardship, and proud to include so eloquent a spokesman for it as Jim Kimmel.

—*Andrew Sansom*
*River Books General Editor*

# PREFACE

The aim of this book is to take you along on a journey of discovery about the San Marcos River and the people who have lived with it. There are several ways to tell this story. Natural scientists might start at the headwaters and go downstream, as the water flows. Historians might deal with events through time for the entire river. Being historical geography, this book combines those perspectives. We work our way downstream, partially because that's the way the water flows, but more because the earliest knowledge we have about the river begins at the headwater springs. But while water flows downstream, time flows in place, so we will sit on the bank and feel the deep flow of time in special places along the river.

The story of the San Marcos River began at least five million years ago. We humans have been here only about twelve thousand of those years. But we are important, at least to ourselves, so a large portion of this book is about the stories of people and the river.

We say "stories" not to imply that the book lacks fact, but to admit that even some fairly recent human history has been lost, or is confused. So this is not a definitive technical history of the San Marcos River and its people. It is the story of a beautiful river, as recorded and told by the wonderful people who live with and love their river.

We express our thanks to the River Systems Institute at Texas State University–San Marcos, John Crane of the Summerlee Foundation, the Cecilia Young Willard Helping Hand Fund, and Texas State University–San Marcos for support for this book. We thank all of the people on the San Marcos River who shared their information, photos, stories, time, and hospitality. Those are many, but several people were especially helpful, including Ron Coley, Doris and Preston Connally, Florence and Tom Doddington, Randy Engelke, Jack Fairchild, Tom Goynes, Jim Green, John Hohn, Martha Nell Holmes, Patsy Kimball, Shirley Rogers Lehman, Glenn Longley, Antoinette May, Scott McGehee, Billy Moore, Jim Pape, Kathryn Rich, Diane and Jason Scull, Gwen Smith, and Duane Tegrotenhuis. Adrienne Booth, Julie Livingston, and Allison Thompson helped with the research, and Karim Aziz prepared the maps. Staff at the San Marcos Public Library, Texas State University–San Marcos Alkek Library, and the Luling Public Library were very

helpful. Britt Bousman, Ernest and Sally Cummings, Al Lowman, Frank de la Teja, and Dianne and Tom Wassenich provided valuable information and constructive reviews. The anonymous reviewers made very useful comments on the draft. Shannon Davies and Jennifer Ann Hobson of Texas A&M University Press and copy editor Mindy Wilson were most helpful and greatly improved the book. We thank landowners who allowed us to photograph from their property, and we thank our friend Kaare Remme for flying us up and down, around and around, to photograph the river.

# CHAPTER 1

# *More than Just a Little River*

$C$LEAR WATER FLOWS FROM the limestone. Clear water that for millions of years has given life to plants and animals, some unique to its springwater. Clear water that slaked the thirst of mammoths and mastodons. Clear water that attracted humans twelve thousand years ago and has held us ever since. Clear water that can cease to flow.

Although the Spanish called it the River of Innocents, the San Marcos River of Central Texas is a deceptive little stream. With most rivers you can draw a line around their watersheds. If water falls on one side of the line, it runs downhill to the river. If it falls to the other side, it goes into the adjacent river—a small-scale idea of the Continental Divide.

But the San Marcos doesn't work that way. At first it seems simpler. It looks like it rises from the magnificent set of springs in San Marcos—clear, strong flowing, and beautiful. But then comes a big rain to the west and the river floods. How can a spring-fed river do that? True, the springs are the source of flow most of the time, but two large, usually dry creeks on the Edwards Plateau can pour huge amounts of water into the upper San Marcos River.

On the other hand, sometimes the San Marcos can flood downstream even if its headwater drainages are not flooding. This is because what we call the San Marcos River below the city of San Marcos really isn't the San Marcos River at all. It should be

*Canoeists paddle on a winter's day on the upper San Marcos with Texas State University's J. C. Kellam building in the background.*

called the Blanco River because the Blanco, considered a tributary of the San Marcos, is the longer of the two. The geographic convention is that the shorter river is a tributary of the longer one. However, this is one of the many places where geography and history collide. The San Marcos was "discovered" and named by Europeans before the Blanco, just as the Mississippi was named before the Missouri. So, technically, New Orleans is located on the Missouri River—so much for geographic convention. At any rate, to draw our line around the full San Marcos watershed, we must include the Blanco River basin as well.

But this still doesn't explain how the San Marcos River springs from the landscape. The springs that supply its constant flow issue from the huge Edwards Aquifer, which is fed by all the rivers and streams that flow off the southeastern edge of the Edwards Plateau, including the Nueces, Frio, Sabinal, and Medina rivers, plus Salado and Helotes creeks, far to the southwest. So if we want to draw a line around the land that contributes flow for the San Marcos River, we end up with a huge area indeed.

This complex nature of the San Marcos River explains why it has been a blessing to the people who have lived with it for twelve thousand years. Even though the San Marcos was deceptive, it was a treasure to the First Peoples and Europeans. Like its sister rivers, the Comal and San Antonio, the San Marcos flowed constantly. Most Texas rivers, while they may not be deceptive, are often cantankerous and uncooperative to humans who want to put them to use. The Brazos and Colorado, big rivers that nineteenth-century settlers expected to use for transportation and waterpower, were often in flood

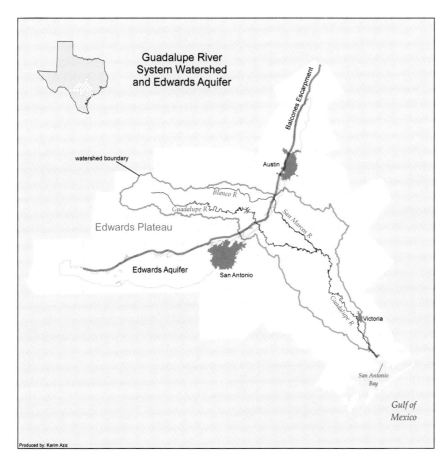

*The headwaters of the San Marcos River were featured on a postcard dated 1906. From the collection of Jerry and Jim Kimmel.*

or drought. Barely useable for transportation and not dependable for waterpower, they would wash away your dam and mill if you tried to put them to honest labor. But the San Marcos, Comal, and Guadalupe rivers gave settlers what they wanted—relatively constant and dependable flow. Anglo American immigrants dammed the rivers, first to turn mills and cotton gins and later turbines to make their towns some of the first in Texas to have electricity.

The constant flow of the San Marcos did not serve only human needs. For the millions of years that the San Marcos Springs have flowed, the river has provided a rich and safe home in which an amazing diversity of life has evolved. Much of the recent history of the San Marcos River has concerned the conflicts between human use of water from the Edwards Aquifer and San Marcos River on one hand and the biological needs of the river on the other hand, which are also greatly important to humans.

## Where and Why?

Probably few travelers on Interstate 35 between San Antonio and Waco wonder why the road follows its particular path, but those travelers are following an ancient trail. It was the route of the railroad. Before that it was a cattle trail to the railroads in Kansas. It was part of the Spanish Camino Real from Mexico City to the colonies in East Texas in the early 1700s. It was an Indian trail, and long before people

*The city of San Marcos and Texas State University originally developed on the slightly higher right bank of the river. The abandoned Gyro Tower, lower right, is on the upper side of the Balcones Fault zone, overlooking Spring Lake.*

came here, it was a trail for buffaloes, mammoths, and mastodons. It was a trail between springs.

The story of why those springs are there and why their flow is so bountiful would be violent if it had happened quickly. The story includes rising and falling sea levels, earthquakes, and even piracy. However, those things happened over a period of about 145 million years, so probably they did not seem violent, or perhaps the dinosaurs and saber-toothed tigers that experienced some of the events recognized the violence as part of life.

The limestone we see around San Marcos Springs was laid down in a succession of seas beginning about 145 million years ago. Perhaps about 100 million years ago, caverns developed in the limestone. Between 24 million to 5 million years ago the western part of the continent, including what is now Central Texas, was uplifted. This uplifting pulled and twisted the land, shattering the old rocks like glass. The resulting major cracks (faults) run in a curve connecting Waco, Austin, San Antonio, and Del Rio, but the rocks between the faults were broken by many small cracks going in all directions. The zone of faulting was along what we now call the Balcones Escarpment, which forms the eastern edge of the Edwards Plateau.

Limestone is resistant to erosion, but it can be dissolved by acidic water. Rainfall and tumbling water in fast-moving streams mix with carbon dioxide in the air, creating a dilute solution of carbonic acid. As this water flowed off the plateau, it entered cracks in the limestone and slowly dissolved it, creating underground cavities. Those cavities hold water, forming what we now call the Edwards Aquifer.

But how is it that water flows out of the aquifer only at certain places? Although the Balcones Fault is about 250 miles long, only twelve large spring systems issue from it. Why are there so few springs, and why are they located where they are?

Geologists Charles Woodruff and Patrick Abbott hypothesized that as the uplift and faulting occurred, the steepening slope in the eastern part of the uplift increased the erosive power of some small streams, which chewed their way up the escarpment, in what is called "headward" erosion. These aggressive little streams eroded their way into the original river basins that flowed eastward off the plateau and captured their flow. This "stream piracy" increased the flow and erosive power of the pirate streams. The geologists figured that the original Blanco River flowed east into Onion Creek and then into the Colorado River, but the pirated Blanco flowed southeast through the channel that is now the headwaters of the San Marcos River. Because this stream flowed from a large area, it was able to erode deeply into the Edwards limestone, cutting into the cavities, thus opening a "drain" in the aquifer. That drain is San Marcos Springs.

Stream piracy probably had a much more profound effect on the landscape and people's lives here than simply creating a big spring. The geologists think that opening the springs increased the capacity of the aquifer. The springs not only allowed water to flow out from the aquifer, but also made it possible for water to flow in from the surface, through cracks and sinkholes in the limestone. Water flowing in from the surface brought a new supply of carbonic acid that dissolved more limestone and increased the size of the cavities that hold water in the aquifer. If the pirate streams had not eroded into the limestone and opened holes into the aquifer, the Edwards Aquifer would not have its current huge capacity that supplies water for more than a million people.

## Spring Flows

The San Marcos Springs consist of three large openings in the Edwards Limestone and more than two hundred smaller outlets. Comal Springs at New Braunfels are the largest in Texas, but San Marcos Springs are second, with an average flow of about 170 cubic feet per second (cfs), which is equivalent to about 110 million gallons per day. The water flows a constant 72° Fahrenheit. In 1691 the Spaniard Terán de los Ríos reported that the Cantona Indians called the springs *Canocanayestatetlo*, which they said meant "warm water."

San Marcos Springs are at the low end of the San Antonio section of the Edwards Aquifer, which stretches from Brackettville to Kyle. This part of the aquifer is about 1,350 feet above sea level at its western end at Brackettville and slopes to 574 feet at San Marcos Springs. The aquifer is about 175 miles long and varies between 5 and 30 miles wide and 400 to 500 feet thick. Since they are at the low end of the pool, the San Marcos Springs would be the last to go dry.

But San Marcos Springs have never gone dry in recorded history. The lowest flow recorded was 46 cubic feet per second—about one-quarter of the average flow—in 1956 at the height of the worst drought on record. Since then the flow has fallen below 60 percent of its average only four times, in 1964, 1984, 1989, and 1996. Flow has exceeded the average in twenty-one of the years between 1957 and 2000.

Use of the Edwards Aquifer affects flow in the San Marcos River. Traditionally in Texas, groundwater belongs to whoever can capture it, called the "right of capture." The one with the most powerful pump gets the water, even if it comes from underneath someone else's land or stops the flow of neighbors' springs. However, the federal Endangered Species Act, administered by the U.S. Fish and Wildlife Service, prohibits harming endangered species such as the five in the San Marcos River. In 1991 the Sierra Club successfully sued the Fish and Wildlife Service for failing to protect the endangered species in the San Marcos River. To avoid federal management of the aquifer, the Texas Legislature created the Edwards Aquifer Authority and gave it power to limit pumping. Regulating a right that has traditionally been free is never popular, and the regulations

*Under the diver are several of the springs that form San Marcos Springs—the source of the river and the site of the Aquarena Springs resort, now the Texas Rivers Center.*

were met with strong opposition that has still not been resolved.

Another controversial aspect of river flow is the historic dam at the headwaters of the San Marcos, just below the springs. The dam was built in 1849 with a removable board at the top to adjust the height of the dam. With the board in place the level of water behind the dam is increased, putting pressure on spring outlets and reducing the flow from the springs into the river. River activists believe that not only the board but the entire dam should be removed.

## River Floods

Contrary to the popular image of Texas as a dry place, it actually rains quite a bit in the San Marcos River basin. The average annual rainfall is 30 inches or greater, which is about what London, England, receives. But there are major differences. London's rainfall is spread fairly evenly throughout the year, while rainfall in the San Marcos River basin tends to come in fast, heavy deluges. For example, the U.S. Geological Survey attempted to measure a storm in late October 1998, but after fifteen hours of torrential rain, its rain gauges overflowed.

The *Hays County Times* of October 2, 1895, described the disastrous results of a rapidly rising river:

> Fentress: The San Marcos put on her war pants [sic] Saturday the 26th and tried to down whom she came in contact with. At 1:15 o'clock she commenced to rise at such a rate that in one hour later she was a vast foaming torrent of drift wood, lumber, mules and stock of all kinds, sweeping by at a terrifying rate.

Imagine that you are floating down the upper San Marcos in one of the bright yellow tubes rented by the Lions Club. The river's 170 cubic feet per second average flow is enough to carry you down the river nicely and provide a little thrill as you pass through rapids or go over Rio Vista Dam. But in 1921 the river flowed 97,000 cubic feet per second at San Marcos—570 times its normal flow. At Luling in the 1998 flood, the river flowed at 206,000 cubic feet per second, which is 524 times the average flow there.

Where does the floodwater come from given that the San Marcos is a spring-fed river? True, the springs provide "base flow." However, the San Marcos also receives runoff from tributaries. Sink Creek, Sessoms Creek, Purgatory Creek, Willow Creek, and York Creek are "intermittent" creeks, meaning they only run when it rains. However, they drain the eastern slope of the Edwards Plateau, which has recorded the highest intensity rainstorms in the continental United States. So when it rains, those creeks dump huge amounts of water into the San Marcos River.

By name at least, the Blanco River is a tributary to the San Marcos. The Blanco's watershed is more than 400 square miles on the eastern portion of the Edwards Plateau. The Blanco carries large floods, running 105,000 cubic feet per second at Kyle in the flood of October 1998. When the Blanco is flooding, its high water acts as a dam that slows the flow out of the San Marcos. The clear springwater from the San Marcos River cannot go into the main flow, so it backs up into the city of San Marcos, causing what people call "clear water" floods. Powerful floods on the Blanco have also made water flow upstream in the San Marcos River. Of course, if both rivers are flooding, there is nothing to do but to seek high ground.

Floods on the San Marcos River have caused grief. Two people drowned in the 1970 flood. A flood in 1981 displaced 1,800 people. These floods motivated the formation of the Upper San Marcos Watershed Reclamation and Flood Control District. With funding from the U.S. Soil Conservation Service (now the Natural Resources Conservation Service), the district constructed three detention dams on Sink Creek and two on Purgatory Creek. The last dam was completed in 1991. The dams reduce floods in two ways. First, they hold the runoff, slowing its flow to the river, thereby reducing the flood level downstream. Second, some of the dams are built over sinkholes in the Edwards Limestone, thus allowing some of the detained floodwater to flow into the aquifer.

*Above, the Blanco River receives the smaller and shorter San Marcos, but loses its name, becoming the San Marcos River. Below, high water on the Blanco has backed up the San Marcos, causing a "clear water" flood on Cape Road at Thompson's Islands.*

The dams served their function in the 1998 flood. Richard Earl, a geographer and hydrologist at Texas State University–San Marcos, calculated that the flow in the San Marcos River through the city was equivalent to a twenty-five-year flood, even though the rain event was the largest that had ever been recorded and far exceeded the projected one-hundred-year flood event. Earl estimated that without the detention dams, the flow through the city of San Marcos would have been in the range of 119,000 to 145,000 cubic feet per second, rather than the estimated 45,000 cubic feet per second that flowed down the upper river. This means that water would have reached the Hays County Courthouse in downtown San Marcos.

The detention dams on Sink Creek and Purgatory Creek have reduced flooding in San Marcos and downstream, but they have also changed the subtle dynamics of the river. A river adjusts its channel to carry high flows that occur roughly every eighteen months. It maintains a channel large enough to hold these flows within its banks. Larger flows go "over bank" into the floodplain. Before the detention dams, the average annual highest flow on the San Marcos in the city of San Marcos was 18,000 cubic feet per second. The detention dams have reduced that to 1,500 cfs, less than one-tenth the normal average high flow. So what does the river do? It adjusts its channel to carry its new most frequent high flow level—it fills in the channel. In the upper part of the river, the channel is one to two feet shallower, partially due to sediment from construction sites that the river has not been able to scour from its channel.

*The limestone Hill Country and its underlying aquifers are the sources of water for the San Marcos Springs. The Blanco River in the foreground is the main "tributary" of the San Marcos. Canyon Lake, in the background, is on the Guadalupe River, which the San Marcos joins near Gonzales.*

## Watershed Characteristics

Although the San Marcos is primarily a spring-fed river, its watershed is the source of floods and has important effects on the river's water quality. The San Marcos is a short river. A fish would swim about 76 miles from the springs to the confluence with the Guadalupe near Gonzales. However, his high-flying cousin the crow would have to fly only about 40 miles to go from one end of the river to the other. But in that short distance the river encounters three ecological regions of Texas—the Edwards Plateau, the Blackland Prairie, and the Post Oak Savannah. The water flowing off each of these regions makes a distinctive contribution to the river.

### EDWARDS PLATEAU

Most of the water in the San Marcos River comes from the Edwards Plateau, either as groundwater from the aquifer or runoff from the Blanco River and other tributaries that drain the eastern portion of the plateau, now called the Balcones Canyonlands. Water from the Edwards Plateau has two special characteristics. First, it is usually clear. The soils of the Edwards Plateau are generally thin, so water running off the landscape does not pick up much soil, and the stream channels are rocky and do not contribute much sediment to running water. Second, water from the Edwards Plateau contains a large amount of calcium dissolved from the limestone by the naturally acidic water.

Water runs off the plateau quickly due to its thin soils and steep slopes. Thus, with the torrential storms that are common to the region and the characteristics of the landscape, the Edwards Plateau can produce large amounts of fast-moving water.

### BLACKLAND PRAIRIE

The San Marcos River flows through the Blackland Prairie from San Marcos to near Fentress. Just as its name implies, the Blackland Prairie consists of deep fine soils with an abundance of organic matter. These soils are easily eroded, so major tributaries such as Plum Creek and York Creek may carry large amounts of soil eroded from fields and pastures and from the banks and beds of the creeks. However, the slopes of the prairie are gentler than those of the plateau. The soil is more absorbent, and there is more vegetation to slow the flow of water across the landscape. Thus, the amount and speed of water coming from this portion of the watershed is generally less than that from the plateau.

Farmers recognized the good soil of the Blackland Prairie during the early Anglo American settlement of the San Marcos basin. Cotton truly was king in the region until the early twentieth century, when the industry was hit by a combination of the boll weevil, pink bollworm, cotton root rot, federal acreage controls, synthetic fibers, and competition from foreign producers and irrigated lands in South and West Texas. The magnificent homes in the historic parts of San Marcos and Lockhart express the cotton wealth. However, cultivation exposes the soil to erosion, and agricultural chemicals are not completely absorbed by the plants. Runoff carries both soil and chemicals to the river.

Much of the Blackland Prairie is now grazing land. Properly managed grazing land holds much of the water that falls on it, thus limiting erosion. However, if the land is overgrazed, erosion can be substantial. Also, cattle that graze along creek banks can damage those areas, and their droppings can add nutrients to the water.

### POST OAK SAVANNAH

From Fentress to where it joins the Guadalupe River, the San Marcos flows through the Post Oak Savannah. This area is more heavily wooded than the Blackland Prairie, sometimes with dense stands of post oak trees. Post oaks are small and somewhat twisty, not good for timber but good for short posts. The trees were prolific, but the demand for oak posts was limited since cedar (juniper) posts were more resistant to decay. Thus, the post oak forests were not cleared, and this part of the San Marcos watershed remains relatively natural.

The woodland and forest landscape of the Post Oak Savannah holds

*The Blackland Prairie produces life's necessities.*

*Although part of the modern world, the San Marcos reminds us that nature is basic to our souls and our existence.*

much of the rainwater that falls on it, limiting the amount of runoff and erosion. Due to the relatively thick forest cover, there was traditionally less cultivation in the Post Oak Savannah than in the Blackland Prairie. Grazing and pecan orchards are now the most common agricultural practices. A substantial amount of the watershed west of Luling contains oil wells, with the potential threat of spills.

The entire watershed has been highly modified by human activity. Native people frequently set fires to aid in hunting. European settlers greatly modified the landscape by clearing, plowing, suppressing natural fires, building cities and towns, extracting minerals such as oil and gravel, and building railroads and highways. Because water runs downhill and carries whatever is loose with it to the river, a river is the product of what goes on in the watershed.

Though pasture is currently the most common use of the watershed, the cities and towns on the river have the strongest impact. San Marcos, Martindale, Staples, Fentress, Prairie Lea, Luling, and Ottine are concentrations of human activity that affect the river with wastewater and pollution from storm water runoff. There are five wastewater outfalls around San Marcos, Martindale, and Luling. Many dwellings use septic tank systems to dispose of wastewater. Most of these rely on the soil to absorb and treat wastewater. These systems can be effective if properly designed and maintained, but often they are old and neglected and may overflow into tributaries of the San Marcos. Pavement and rooftops do not absorb water as the natural landscape does, and consequently they increase runoff. City, county, state, and federal agencies all have responsibilities to protect the quality of the San Marcos River. However, as the population of the region continues to grow, threats to the river will increase.

## Water Quality

The special features that have made the San Marcos River attractive through time—its clarity and the quality of its habitat—all depend on water that has a high oxygen content and is relatively free of nutrients, sediment, pathogens, and toxic materials. That is a tall order for a river in the modern world.

The San Marcos River has traditionally had high quality water because most of it comes from the Edwards Aquifer. We generally think of springwater as pure because most aquifers exist in sand or sandstone that filters the water as it passes through. Not so for the Edwards Aquifer. Because it is recharged by water running directly into cracks and sinkholes in the limestone, it does not filter the water as well as a sandstone aquifer. It is extremely vulnerable to pollution in the contributing watersheds and over the recharge zone. The state of Texas and the city of San Marcos have regulations aimed at preventing pollution of the aquifer, but the risks increase as the Hill Country west of San Marcos continues to be developed for residential use.

# CHAPTER 2
## *Life of the River*

WE OFTEN PUT our kayaks in the river at Rio Vista Park in San Marcos and paddle about a mile upstream, until we come to the 1849 dam that forms Spring Lake. Then we lie back and let the kayaks drift with the swift current. Because we are quiet we see kingfishers and herons, turtles and snakes, gar and bass. Although the river passes through the center of San Marcos, there are stretches where we can imagine ourselves on a remote semitropical river, lush with vegetation on the banks and in the clear water. The wild cry of a red-shouldered hawk confirms that we are in a place where nature is prolific.

The long-term stability of the San Marcos River provides a riparian habitat that supports an especially rich community of plant and animal life, including ourselves. Water, the main feature of a river ecosystem, supports plant life, which in turn provides the habitat for all other forms of life. The profusion and diversity of aquatic vegetation in the San Marcos River are fundamental to the character of the river.

As the river flows downstream, many aspects of its life remain constant, but important differences begin to emerge. From the headwater springs to its junction with the Blanco River, the San Marcos' spring-fed flow is clear and has a constant

## River People

Untold thousands of people use and enjoy the San Marcos River, but a few have literally dedicated their lives to her. They know the river intimately and speak for her when she is threatened.

Tom Goynes and Duane Tegrotenhuis and their families live on the river and make their livings from it, with their campground and canoe livery, but the river has given meaning and purpose to their lives for many years. Tom first ran the Texas Water Safari in 1967 but pulled his battered wood and canvas canoe out at Luling, short of the finish line. The next year he completed this 260-mile race, which ends at Seadrift on San Antonio Bay. He won it several times in succeeding years. Tom moved to San Marcos from Houston in 1972, planning to run a campground and canoe shop to pay for law school. The river won out, so the San Marcos has a dedicated proponent, and the world has one less lawyer.

Duane Tegrotenhuis came from Iowa, seeking warmth in the South. He bought Tom Goynes's canoe shop and spends his days helping people enjoy the San Marcos River. He says that even after twenty years, especially after all that time, living with the river is still a profound experience for him.

Jack Fairchild, an aeronautical engineer who grew up in Houston, spent his summers as a youth in the 1930s and 1940s staying in a cabin his family leased each summer at Rogers Park. His boyhood love of the river brought him back to spend his "retired" years as a tireless and scientifically astute leader of the San Marcos River Foundation.

Dianne and Tom Wassenich lived on and with the river at Martindale. As they saw it deteriorate from upstream pollution, they became active in the San Marcos River Foundation. Dianne now serves as executive director. Tom sold his successful restaurant and completed a graduate degree in geography at Texas State University–San Marcos, building on the expertise he had developed regarding environmental flows in Texas rivers.

Ron Coley got a scholarship from Jacques Cousteau's organization to learn to be an underwater cinematographer in the early 1980s. Rather than the marine films that made Cousteau famous, Ron was advised to do a film about freshwater. He teamed up with a producer from the British Broadcasting Corporation and made *The River of Innocence* in 1983, a film that tells a poignant story of the river. Ron continues to tell the stories of the upper river as director of the Aquarena Center.

Randy Engelke grew up outside Luling, got a degree in agriculture from Texas A&M University, worked in the Luling oil field, and then found his passion as the parks and recreation director in Luling. Randy has been the quiet force behind the restoration of Zedler's Mill in Luling. He is passionate about old mills and wants to preserve the remaining ones on the San Marcos to tell the story of this crucial part of the river's history.

---

cool temperature, low nutrient levels, and a fairly constant flow. The Blanco brings in warmer water with more nutrients. As the river flows downstream small tributaries add nutrients and fine sediments from the deep erosive soils of the Blackland Prairie and the Post Oak Savannah.

The river moves more slowly as it flows across the coastal plain. Warmth from the sun and nutrients from the soil give life to a variety of single-celled organisms, called plankton, suspended in the water. The sediment and the plankton cloud the water, giving it a pea-green color and preventing light from penetrating into the deeper water as it does in the upper river. The rich mixture of submerged vegetation of the upper river cannot live here. The tall canopy of trees shades the river in the summer—in some places it is almost impossible to see the river from the air.

*The dependable flow and constant temperature of the upper river provide habitats for a great variety of native and nonnative plants and animals, including the beautiful but highly invasive water hyacinth.*

*Downstream the river receives runoff from the surrounding land, becoming less clear. The riparian area is heavily wooded for the entire river's length, as it is here at Fentress.*

We know more about the life in the upper river than the rest of the river. The upper river is pretty, accessible, and fascinating to biologists because of the great diversity of life in the clear water. As with many things, beauty attracts attention.

The biology of the San Marcos River is of great concern because it is changing. At least one native species has probably become extinct, and four others are endangered with extinction. Other native species are crowded by aggressive newcomers. Let's consider first the natives and then the nonnatives.

### Native Species of the San Marcos River

Biologists at Texas State University–San Marcos, the University of Texas–Austin, the Texas Parks and Wildlife Department, and the U.S. Fish and Wildlife Service have observed and described twenty-one plant species, not including algae, and forty fish species in the San Marcos River. In addition, the river corridor supports a great variety of floodplain vegetation and the animals that are attracted to that rich riparian habitat. (See appendix 1 for a description of native species in the San Marcos River.)

The riparian corridor hosts numerous tree species, including pecan, American elm, cedar elm, cottonwood, sycamore, black willow, live oak, Carolina ash, burr oak, water oak, box elder, walnut, and baldcypress. Of these, baldcypress is especially important and interesting.

### BALDCYPRESS

Float down the river and pick out the most outstanding tree. Your choice will probably be the very tall trees with straight trunks and feathery leaves, growing with their roots in the water. Baldcypress (*Taxodium distichum*) grows in saturated soils and can withstand deep floodwater flows up to 4 miles per hour. These trees naturally occur along the Atlantic Coastal Plain from southern Delaware to southern Florida and westward along the Gulf Coast into Texas. They grow up the Mississippi Valley as far as southwestern Indiana and follow streambeds onto the Edwards Plateau, which is their westernmost range. Baldcypress trees in virgin forests in the southeast wetlands grew as tall as 150 feet. The "champion" baldcypress, in Louisiana, reached a diameter of seventeen feet.

*Water from the river moves into soil pores along the banks, supporting profuse vegetation.*

*Texas' most common turtle, the red-eared slider, cold-blooded like all turtles, spends much of its time in the sun. Elephant ears and other exotic vegetation are common in the San Marcos.*

Heartwood from old-growth baldcypress trees contains oil that makes the wood resistant to decay. Lumber from baldcypress trees was an important building material in early Texas, especially because it was easily split into flat thin "shakes" for roofing shingles. However, wood from younger second-growth trees does not resist decay.

To help withstand the power of floods, baldcypress trees put out extensive root systems. The intertwined roots make a natural bulkhead that protects the riverbank from erosion. The roots and buttressed base also make baldcypress trees highly resistant to wind—they can usually withstand hurricane force winds. Baldcypress roots often put out vertical conical extensions called "knees." Cypress knees may be as tall as twelve feet, but their function is still not clearly understood.

### EELS, PRAWNS, SWAMP RABBITS, AND WATER BIRDS

The water and rich vegetation of the San Marcos River provide habitat for a variety of animal life. There are three wonderfully weird and fantastic animals associated with the San Marcos River that you may never see. One is a far-traveling eel; one looks like a crayfish but is the size of a lobster; and the other is an underwater bunny. Along with these oddities is a wide variety of birds, especially ducks, large waterfowl, and hawks, all of which are attracted by the river's protected habitat and the variety of plant and animal life there.

## American Eel

Although thousands of people come to the San Marcos River to swim and play, surely none makes the voyage the American eel does. Eel larvae come from the Sargasso Sea in the western Atlantic Ocean. Biologists assume the adult eels spawn there, though they have never found adults there. Hatched in January through March, the clear, leaf-like larvae, about two and a half inches long, develop into transparent "glass" eels as they travel to freshwater. Once in freshwater, the eels turn yellow to olive. They remain in freshwater, feeding on snails and small fish for the next seven to thirty years. When they reach sexual maturity they turn bronze or black with a silver sheen. They then retrace their voyages to their spawning area, whether the Sargasso Sea or some yet undiscovered place.

Eels can travel overland if necessary. Since they can absorb oxygen through their skin, they can live out of water for some time, but they must remain in moist conditions. To seek suitable habitat, an eel will slither overland, laying down a film of slime. When that eel has exhausted itself and its slime, another will follow and extend the path. People have also

*The lady fishing at the head of Spring Lake was dressed in the formal style of the early 1900s. From the postcard collection of Jim Pape.*

seen intertwined eels roll up beaches searching for freshwater. Eels, popular food in Japan and Europe, can grow to more than 4 feet long and up to 10 pounds.

## Giant Freshwater Prawn

The giant freshwater prawn, or big claw river shrimp (*Macrobrachium carcinus*), can be a foot long and weigh more than 2 pounds, though the occasional snorkelers in the river who see one always claim they are even bigger. The giant freshwater prawn is found along the Gulf Coast to Mexico and in the Caribbean and Brazil. They were so plentiful in the 1880s that commercial fishermen caught and sold them. Sadly, the many dams on Texas rivers, including the Guadalupe and San Marcos, have made their lives a difficult obstacle course, and their numbers have greatly diminished.

Giant freshwater prawns have a complex life cycle that relies on both the clear freshwater of the upper San Marcos River and the brackish water of the Guadalupe River's estuary. Adults spend their lives in freshwater, but their larvae must quickly travel downstream to the estuary, where they mature. The new adults then travel up the river. Our dams, however, make both of these trips extremely difficult.

The prawn larva from the upper San Marcos, now a young adult driven by its small brain to return to those springwaters it somehow knows, must travel 200 miles upstream and go over eight dams. The trip may take two years, but instinct and the imperative to breed push the

*Life of the River*

*The magnificent baldcypress trees protect the river bank from erosion but also remove portions of the bank when they die and fall.*

young prawn upstream. The dams built in the 1800s slowed the great shrimp, but since water flowed over the tops of the dams, the persistent young shrimp were still able to make their way upstream. Fortunately, all of the dams on the San Marcos and Guadalupe are the old overflow dams. Unfortunately, the new dams, which release water from the bottom, are impassable barriers for the prawns. The dam planned for the Guadalupe River near Cuero will block the prawn's upstream pilgrimage.

Even the old dams affected the life cycle by slowing the trip downstream for the newly hatched larvae. They die if they don't make it to the brackish water within a few days of hatching. Before the dams, the larvae could float down the river to arrive at the estuary in time, or they could catch a flood and make it with time to spare. But the dams greatly slow the flow, thus increasing the mortality rate of larvae.

*Swamp Rabbits*

Many San Marcos characters call themselves "river rats." Even a local Internet company offered that name as an e-mail address. But not many people call themselves "swamp rabbits," perhaps because they don't know about these water-loving bunnies that come out only at night.

Swamp rabbits confirm three stereotypes. First, they breed like the proverbial rabbit. They can produce two or more litters per year of up to six bunnies each. That's maybe as many as eighteen offspring per female each year. Second, they are fast, running zigzags at almost 50 miles per hour. Third, like Brer Rabbit, they seek the briar patch, rather than the burrow, for protection. Swamp rabbits have a well-defined territory that they will not leave, even when chased by dogs. They know that a dog will not follow them into the briar patch.

While swamp rabbits may live up to certain stereotypes, their scientific name, *Sylvilagus aquaticus,* tells us they are unique. They go into the water. Their thick fur is like the modern diver's dry suit, allowing the swamp rabbit to swim and even to hide out underwater with just its nose exposed. This is probably good protection from the coyotes, hawks, owls, and people that prey on the swamp rabbit, but it is not an effective strategy for hiding from its other major threat, the alligator. Though the briar patch sounds better and better, even it was not able to keep humans from severely depleting these rabbit populations. Six pounds of rabbit was worth some serious chasing, even into the briar patch.

*Birds*

The San Marcos River is included in the ranges of at least twenty-nine bird species that depend on river habitats. Ducks, of course, are most common, and account for eleven of those species, though perhaps the most exciting are the huge great blue herons and the wild and free red-shouldered hawks.

If you truly love the river you might consider life as a duck—they seem to have the best of all worlds. Imagine floating on the river all day, quacking with your friends. When you're hungry you dive into the clear water and find fresh food. And you can fly. Few animals have the three-dimensional world of a duck—a swimmer, diver, flyer. They aren't graceful walkers, but they seem proud of themselves nevertheless.

Although ducks have enviable lives, and some are quite handsome, they are not what most people would consider majestic. But a majestic, regal bird found on the river is the great blue heron, which stands more than 4 feet tall as it stalks along the shallow edge of the river watching for fish and frogs. As we glide close in our kayaks, the heron slowly flaps away on its 6-foot wingspread, its long neck folded in an S-shape. If the bird is sufficiently irritated by the disturbance, it may leave with a prehistoric-sounding squawk.

Living in modern San Marcos, you might think that the wild is gone. The interstate roars and the rock concert at Sewell Park blasts. But then you hear an ancient wild cry from the sky and see two red-shouldered hawks flying the thermals that rise with the escarpment above the springs. These hawks are especially adapted for life in the riparian forest, with extra long tails that allow them to twist and turn

*The ancient-looking great blue heron is an attentive fisher.*

through the trees. They feed on small birds, rodents, snakes, and lizards, even landing on our deck on the hill above the springs to lunch on anole lizards.

## OTTINE SWAMP

About 50 river miles downstream from the headwater springs, near Ottine, is a part of the San Marcos River that is biologically unique. Called by locals the "Ottine Swamp," it was incorporated into a state park in 1938 with the more refined name of Palmetto State Park, for the dwarf palmetto plants that grow there in profusion. The San Marcos River usually flows slowly through this region, its water pale green due to the naturally rich nutrient content and plankton growth.

The river has meandered through the area, leaving an oxbow lake and lowlands that are frequently flooded. Floods, plus the high water table, supported luxurious plant growth more typical of that farther east. Oil pumping in the 1930s lowered the water table so that plants now depend on water pumped from the aquifer by an antique hydraulic ram pump. Unfortunately, the wetland is subject to droughts that will ultimately eliminate the species that depend on high water levels.

The most unusual plant in the wetland is the palmetto (*Sabal minor*). Its trunk is underground, and the leaves may be up to 5 feet in length. Palmettos are one of the heartiest palms in the world. They can resist freezing temperatures that would kill other palms, but they also require hot summers.

## ENDANGERED AND THREATENED SPECIES

The unique character of the habitat provided by the San Marcos River has resulted in the evolution of life forms found only in its waters. The habitat of these unique species is rapidly changing, and the U.S. Fish and Wildlife Service has designated four

*The little palms of Palmetto State Park heighten the sense of mystery in this important wetland.*

species in the San Marcos as threatened or endangered. Unfortunately, one other species is now considered extinct, but might yet be found.

*San Marcos Salamander* This small amphibian with feathery gills is found in Spring Lake and the area immediately below its dam. Classified as threatened, it is affected by reduction of spring flow, habitat modification, and reduction of water quality.

*Texas Blind Salamander* This endangered salamander lives in the perpetually dark recesses of the Edwards Aquifer underneath the city of San Marcos. It is at risk due to diminished spring flows and groundwater pollution. It is seen only when flow from the aquifer or pumping brings it to the surface.

*Fountain Darter* This minnow-like perch, a diminutive relative of the walleye, is endangered and is found only in the upper 3 miles of the San Marcos River and the Comal River in New Braunfels. The fountain darter was first collected in 1884 when it was abundant in both rivers. The fountain darter is threatened by habitat loss, cessation of spring flow, siltation, reduction in water quality, river channel changes, and introduction of exotic species.

*Texas Wildrice* This aquatic grass species is endangered. Its long, narrow, flowing leaves may be as long as 4 feet, and its long stems may reach 12 feet. It is found only in the upper mile and a half of the San Marcos River. It is rooted to the bottom in water 1 to 6 feet deep. Texas wildrice is related to wildrice that grows in the central, northern, and eastern U.S. but not elsewhere in Texas. In the 1930s the grass was so thick that it was mowed in the river to maintain swimming areas and flow in irrigation canals. The Texas wildrice population declined about 40 percent after the 1996 drought and the 1998 flood. Threats to Texas wildrice include cessation of spring flow, recreation, decreased water quality, grazing by introduced species, and the effects of damming the stream.

*San Marcos Gambusia* This small mosquito fish was found only in still waters in the upper San Marcos River. In 1982 specimens were taken from the river with the hope of repopulating and thus saving the threatened species. However, the specimens were all found to be genetically contaminated through hybridization with the more abundant and widespread western mosquito fish. The San Marcos Gambusia was last seen in the river by biologists in 1983 and declared extinct in 1984.

## Nonnative Species in the San Marcos River

Humans have introduced many plants and animals to the San Marcos River in the past one hundred years, and some of them are drastically affecting the river ecosystem. Biologists have identified sixteen nonnative plants and eighteen nonnative fish in the river. (See appendix 2 for a description of nonnative species in the San Marcos River.)

Nonnative species can seriously affect a river's ecology. They are often "generalist" species that can live successfully in a wide variety of habitats. As habitats become modified, the nonnatives can push out native species that are more specialized for the natural conditions, thus reducing biological diversity. However, as is often the case with foreigners, some of them have fascinating stories. For example, the elephant ear plants that line the banks of the upper San Marcos are of the same genus as the plant that is called the "potato of the tropics." About 100 million people in Southeast Asia, the Pacific Basin, Africa, Egypt, the West Indies, and South America rely on these food plants known variously as taro, dasheen, or cocoyam. The plants produce a starchy underground stem that can grow to more than a foot long. Taro has more protein than Irish potatoes and is particularly easy to digest. However, taro must be boiled for an hour or so to remove needle-like crystals of calcium oxalate, which make taro virtually poisonous. The Texas Agricultural Experiment Station at Angleton experimented with elephant ears between 1914 and 1924 as a possible replacement for Irish potatoes. Their experiment was not a success.

While elephant ears may be the most visible exotic plant in the upper river, they are not necessarily the most

*The endangered Texas wildrice grows only in the upper San Marcos River.*

harmful. That accolade may go to the water hyacinth. The water hyacinth, with its pretty blue-purple flowers, is one of the most productive water plants in the world, and considered to be the worst of the invaders. It is a detriment to rivers and lakes because its extensive growth shades out native plants. Also, the water hyacinth floats on the surface, and low oxygen conditions may develop under the mats. Its leaves stand erect and serve as sails. Winds and currents move these floating plants into huge dense mats. We have towed small mats out of the main channel with our kayaks—quite a job. The mats hinder recreation and can block water intakes.

The water hyacinth was introduced to the United States from South America, probably as an attraction in the World's Industrial and Cotton Centennial Exposition of 1884–85 in Louisiana. It has spread throughout the Gulf states. It has been controlled in Florida with chemical herbicides, but the sensitive environment of the upper San Marcos River is too vulnerable to use chemicals here. The San Marcos River Foundation has an ongoing campaign to remove water hyacinths by hand, but it is a slow battle.

A new invasive plant has recently been found in the upper river, the watertrumpet (*Cryptocoryne beckettii*). Native to southeastern Asia, watertrumpet is a popular aquarium plant. Its colonies spread from bank to bank and exclude native species. Herbicides may kill it, but they endanger native species. Digging it out is ineffective unless the entire tuberous root stock and underground stem are removed.

The Florida red-bellied turtle (*Pseudemys nelsoni*), native to Florida and southeastern Georgia, is the only nonnative reptile found in the river. Its population is well established but localized to Spring Lake.

The San Marcos River is now home to an exotic clam and three nonnative snails. The Asian clam (*Corbicula fluminea*) is established and common in the river. Biologists David and Beth Bowles have observed that shells from these clams cover much of the surface on parts of the river bottom. The red-rimmed melania snail (*Melanoides tuberculatus*), from southern Asia, is established but not common in the river. It serves as a host for flukes that are parasites in the endangered fountain darter, in waterfowl, and in humans. The quilted melania snail (*Tarebia granifera*), also from Asia, now makes up about 60 percent of the bottom-dwelling invertebrate community in parts of the river and is displacing native snails. It is also a host to the human lung fluke. The most dramatic exotic mollusk in the San Marcos is the giant rams-horn snail (*Marisa cornuarietis*), from South America. This snail grows to almost an inch in diameter.

The almost beaver-sized nutria is the dominant mammal of the river, although it is native to South America. Nutria were introduced to North America to control aquatic vegetation. Unfortunately, their very high reproductive capacity and the lack of natural predators quickly led to overpopulation. Nutria eat the same aquatic vegetation as many waterfowl and thus reduce the waterfowl populations. Nutria are raised in South America for their fur, which is sold in Europe. However, there is no demand for nutria fur in the United States.

The appearance of many nonnative birds near the river is no surprise. Storms carry birds far from their native range, and people introduce birds as curiosities or to raise them for meat and feathers. Careful observers have seen a variety of exotic birds on the river, including black swans and geese from Egypt, China, Africa, and Canada. Perhaps the most outstanding nonnative bird on the upper river is the stately European mute swan. These birds are especially associated with England, where rivers and their birds are indeed taken seriously. Mute swans were given royal status in the twelfth century. Owners marked their swans on the beak with officially registered marks. Each summer, at the swan "upping," owners marked the young cygnets with their parent's marks—just like a Texas cattle roundup.

The English kept their mute swans for prestige and food, but the Greeks had more imagination. Greek gods could turn themselves into swans for seduction. Zeus transformed himself into a swan and visited Leda, wife of the King of Sparta, on her wedding

*The San Marcos River is much loved by its family of volunteers who work to clean it and remove nonnative vegetation.*

*Native, nonnative, and domestic birds gather at the river.*

night. Pollux was the result of that union. The mute swans of the upper San Marcos are apparently more sedate or at least more circumspect, as there are no similar legends here.

The San Marcos is much more than just a little river. It is an ecological treasure, but one that is threatened both by our attraction to it and by our negligence.

# CHAPTER 3

# From the Deep Past at the Springs

*I*MAGINE YOURSELF twelve thousand years ago, walking with the other twenty members of your clan across the prairie that we now call Central Texas. It's in the last moon before the cool season begins. The air is hot and dry. You and your people have eaten only a few field mice lately, and everyone's bellies are tight with hunger. You are heading toward the setting sun, looking forward to a little cool in the night but knowing the biting flies will make it hard to sleep. You see a line of low hills on the horizon and decide to walk there to make camp.

As you approach from the northeast, you see a profusion of trees. When you get closer the air seems to change. It's a little cooler, and you smell water. The five-year-old boy and seven-year-old girl break and run toward the trees, even though their mothers yell at them to stop, and run after them.

When you reach the water you see the children and their mothers splashing in a broad shallow pool. Tall cypress trees grow out of the shallow water along with reeds. The water is as clear as the air. The children and their mothers chase fish, then run after the frogs that croak and jump as people walk along the water's edge.

You and your people have never seen a place like this. Water gushes from the surface as high as a grown man. There is precious water to drink and a world of things to eat. Maybe

*These baldcypress trees are perhaps hundreds of years old, and they descend from trees that were present when the earliest people reached the springs.*

## Timeline of Prehistoric People at San Marcos Springs

| Period | Time Span | Lifestyle |
| --- | --- | --- |
| Paleo-Indian | 11,500 Before Present to 8,800 BP | Nomadic bands of twenty-five to fifty people preyed on herds of mammoths, bison, mastodons, and horses. They also ate turtles and tortoises, badgers, raccoons, mice, and alligators. They made sophisticated flaked projectile points from chert and obsidian that tipped their spears. They may have thrown large darts with an atlatl. These points, first found at Clovis and Folsom, New Mexico, are found throughout North America, indicating a widespread cultural diffusion of these people. |
| Early Archaic | 8,800 BP to 6,000 BP | A climatic change resulted in higher temperatures and a drought that lasted for two thousand years. The large animals were now extinct. In the area of the springs, people's diets shifted to deer, fish, rodents, rabbits, and plants, including prickly pear. The overall population was smaller than in the earlier times when the big animals provided sustenance. |
| Middle Archaic | 6,000 BP to 4,000 BP | This is the period of greatest drought and most limited food supply. People ate more plant foods, including nuts, some acorns, and camass, of the lily family. After the drought peaked about 5,000 BP, oaks began to spread across the Edwards Plateau, and pecans flourished in the river floodplains. Both trees provided valuable food. People continued to eat deer and rabbits and perhaps an occasional bison. |
| Late Archaic | 4,000 BP to 1,200 BP | The diet shifted away from acorns and became more diverse. Bison disappeared from the region about 1,200 BP. |
| Late Prehistoric | 1,200 BP to 475 BP | People made a significant technology shift from the atlatl to the bow. Bison returned to the region about 650 BP. At the same time, people in the springs area began to make ceramics. The prehistoric period ended as Europeans entered and began to influence the region. |

your people don't yet have an idea of heaven, but you know that you have arrived at a very special place indeed.

People have come to these springs for at least twelve thousand years. This one place, this bountiful flow of water and life, has been not just the stage, but a player in the lives of people since the deep past.

Archaeologists divide the local prehistoric record into five time periods, based on characteristics of technology and evidences of lifestyle. Those time periods are the Paleo-Indian, Early Archaic, Middle Archaic, Late Archaic, and Late Prehistoric. Of course, archaeologists disagree about some of the particulars, but the broad pattern can help us understand the lives of people at the springs.

The archaeological record at San Marcos Springs tells us little about the life of the people thousands of years ago except that they were there. Dr. Joel Shiner, the Southern Methodist University archaeologist who first investigated the springs area, speculated that the place was so attractive that people lived there continuously. However, this goes against the current understanding of the lifeways of these ancient people. Most of the people during the prehistoric period were nomads who moved frequently to find food. Even with the biological richness of the spring pool and the river, populations of animals and plants would have eventually been decimated if any substantial number of people had lived there permanently.

The springs area was a pleasant place, but life there for humans was hard, dangerous, and short. They commonly ate seeds, roots, and fruits. They also preyed on weak or ill animals and ate insects. But the mainstay of life for the earliest Paleo-Indian people was the mammoths, mastodons, and bison that provided food and materials for clothing, shelter, and tools.

Although the huge animals were dangerous prey, the earliest people may have had a relatively pleasant life. For much of the time the climate was cooler and wetter than it is now. Because the big animals provided such a rich resource, active adults probably worked about twenty hours per week. Archaeologists have found no evidence of conflict such as skeletons with Clovis points in vital parts. Possibly the resources were so abundant and the people so few that there was little need for competition and conflict. More likely, we just haven't found the remains.

But life was fragile. A minor injury or infection would often lead to death. Childbirth was a highly risky experience. The big animals that supported life—huge lions, saber-toothed cats, and strong-jawed dire wolves—were also a major risk to life. Because humans were newcomers, these animals may have had no fear of them and consequently may have been more aggressive. But compared with times to come, the life of Paleo-Indians was pleasant.

Virtually all of the big animals were extinct by ten thousand years ago, and the quality of life deteriorated rapidly as a result. Although deer may have been plentiful most of the time, they were harder to approach for the kill than the mammoths, and one deer did not supply much food or material. So people had to spend more time hunting and processing their kills. Because deer populations are directly affected by drought, sometimes they were not abundant, which forced people to eat rodents, rabbits, fish, and probably anything else they could catch. Surely at times such as these, the springs area was even more important than ever because it provided habitats for a variety of small animals and plants. And the springs were a dependable source of water.

The region's climate has varied substantially through time, but about seven thousand years ago the climate became hotter and drier, much drier. This drought was not the seven-year kind that we think of as normal in Central Texas. This drought lasted for two thousand years. The drought peaked about five thousand years ago, and life slowly changed again for the better. The people of the Middle Archaic period took advantage of the increasing number of oaks on the plateau to the west of the springs and the pecan trees in the river corridor.

While modern people collect pecans from the trees near today's springs, few would consider eating an acorn. However, acorns provide protein, carbohydrates, and calcium, but

they must be processed to remove the tannic acid that makes them too bitter to eat right off the tree. Tannic acid can be removed by boiling, or the acorns can be soaked in a fast-moving stream for several weeks. The meat of acorns can be deep fried. Acorns can also be dried or roasted and ground into meal. The meal can be used to make bread, mush, or pancakes.

*Archaeologists have found a variety of Paleo-Indian and Early Archaic projectiles in the area of the springs.*

## Mammoth Hunt

Imagine killing a fourteen thousand-pound mammoth with spears. The animal stands twelve feet at the shoulder and can kill you with one step or a sweep of its head. You hide in the low brush beside the spring pool, waiting for the animals to come to water. If you are using a spear you must wait until the mammoth is within a few feet of you, then step up beside it and shove the spear into its entrails.

You and your partners probably cannot kill the huge mammoth with your first attack. You will wound it and then chase it, continuing to attack the enraged animal until it bleeds to death. You and the mammoth are both fighting for your lives. Both of you may lose.

But when the kill was made, there must have been rejoicing. There would be thousands of pounds of meat and a hide that would provide clothes for the whole group. Nothing would be wasted. The sinew, intestines, brain, and the ivory tusks were all useful. The fresh bone could be flaked like stone to make cutting tools. While nothing is known of these early people's ability to preserve meat, given the sophistication of their hunting weapons, they had probably also developed the technology to dry meat.

---

Some archaeologists speculate that the work and cooperation required for processing acorns led to a new level of social organization among the scattered bands. The plateau live oak that is abundant today above the springs was probably as abundant five thousand years ago, if not more so due to fires more common then that would have suppressed juniper and allowed more oaks to grow. The increasing abundance of oaks may have decreased the importance of the springs as a food resource during this period.

Although we don't know this from San Marcos Springs, people of this period fished with hooks and lines, nets, traps, and spears. In addition to fish, they ate crawfish, waterfowl, turtles and tortoises, and alligators. They also killed a variety of animals including bears, deer, opossums, squirrels, raccoon, turkeys, and snakes. Edible plants were abundant, including plums, prickly pear, blackberries, and of course, acorns. Once again, the springwater and the pool it formed would be the people's supermarket.

Until about 1,200 years ago the people who spent time at the springs lacked two of the technologies we associate with "Indians." They did not use bows and arrows nor did they make pottery. Their main hunting tools were the spear, the dart-throwing atlatl, and traps. Instead of pottery they used vessels made of animal hides or stomachs or woven baskets. Obviously, these could not be heated over a fire for cooking. They used rock-lined fire pits for cooking and placed heated rocks in the hide vessels or baskets to boil water.

About 1,200 years ago hunters in the region of the springs began to use bows with light arrows rather than spears and atlatls with their heavy darts. Modern atlatl enthusiasts argue its benefits, but the bow and arrow is probably more consistently accurate. More importantly, it is easier to use a bow and arrow from a hiding place in the brush. Drawing a bow requires very little movement compared to the long overhand swing of the atlatl. Deer were probably startled by the throwing motion of the atlatl and moved before the dart reached them. But an arrow could be on its way with little visible motion. The hunter could easily carry several arrows and quickly "reload" and shoot multiple times.

Apparently the benefits of the bow and arrow were not limited to killing whitetail deer. This is the first period in prehistory in which humans were buried with arrow points in vital places. Life around the springs took on a serious new threat.

But the times also brought pottery, a new technology that radically

*The springs area was rich with things to eat, many simply for the gathering.*

*Early hunters may have used a stick, now called an atlatl, to increase the leverage of their throw. Drawing by Jerry Touchstone Kimmel.*

changed the lives of women. Pottery probably affected prehistoric women's work in the same way that electricity revolutionized life for women in the early twentieth century. Suddenly their work became easier and required less time. Until they had ceramic vessels, women had to gather fuel and stones, build a fire to heat the stones, and then place the stones in a skin vessel or basket to boil acorn meal or meat. They also had to make more trips to the water source because their water vessels leaked. After the invention of pottery, women could build a fire and put on the pot. Of course, they still had to catch or gather the food and prepare it, as well as make the pots. Women's work remained much like this in the area until the flow of the springs was harnessed to generate electricity that could pump and heat water.

Like our dependence on electricity, this new pottery technology came with a cost. Pottery pots are heavy and fragile. Since the people were still nomadic, moving became more burdensome, but some archaeologists think they hid the pots at sites and retrieved them when they returned. Also, someone had to develop the specialized skill to make pots. Those who did not have the skill or did other things had to trade something for the pots. Life was easier than the good old days, but not as simple.

Although life changed during the first eleven thousand years that people lived near the springs, the changes were slow and subtle, taking place over many generations. Probably few individuals experienced any substantial change in lifeways compared to their parents or grandparents. However, about five hundred years ago, as they camped and traded at the springs, travelers from the south told of things that the people could not believe.

They told of men who came on huge winged canoes, wore shiny clothes that deflected arrows, rode beasts, and brought death. Seven hundred and fifty miles southwest of the San Marcos Springs, Hernán Cortés invaded in 1519 and destroyed the powerful and complex Aztec society that flourished there. Even though people who visited the springs may have heard these stories for years, they probably had no firsthand experience with Europeans until 1528, when Cabeza de Vaca was shipwrecked near Galveston. Cabeza de Vaca wandered and traded through much of south Texas, finally making his way back to Mexico City in 1536. He left no indication that he visited the springs, but we can imagine that people at the springs heard about him. The stories about strange men grew more real and probably more fantastic, since Cabeza de Vaca was sometimes considered a healer. Even so, he reported that half the native population died of a stomach illness after he and his companions came ashore. The relative stability of the past 11,500 years began to change—with unaccustomed speed.

In 1542 a remnant of De Soto's expedition led by Luis de Moscoso Alvarado reached a river flowing out of "mountains." Some say this expedition only went as far west as the Trinity River or perhaps the Navasota River in east central Texas, but more recent analysis indicates that they may have reached the Guadalupe River. If so, the native people they saw were probably typical of people who might have been at the San Marcos Springs as well. Their chronicler stated:

> There the Indians told them that ten days' journey thence toward the west was a river called Dayco where they sometimes went to hunt in the mountains and kill deer; and that on the other side of it they had seen people, but did not know what village it was … after marching for ten days through an unpeopled region reached the river of which the Indians had spoken. Ten on horse, whom the governor had sent on ahead, crossed over to the other side, and went along the road leading to the river. They came upon an encampment of Indians who were living in very small huts. As soon as they saw them, they took flight, abandoning their possessions, all of which were wretchedness and poverty. The land was so poor, that among them all, they did not find an "alqueire" of maize.

Incredibly, it was almost 150 years before the Spanish came back into what is now Texas, and their return was motivated by fear that the French

## Timeline of Spanish at San Marcos Springs

**Early 1500s,** stories about Cabeza de Vaca probably were told at the springs.

**1691,** Domingo Terán de los Ríos and Father Damián Massanet camped at San Marcos Springs and met Cantona Indians, who called the springs *Canocanayestatetlo,* meaning "warm water."

**1693,** Governor Gregorio de Salinas Varona stopped at the springs on his way to the East Texas mission, establishing the springs as a point on *El Camino Real.*

**1709,** Capitan Pedro de Aguirre and Father Isidro Félix Espinosa stopped at the springs, and Espinosa wrote a glowing description of the country.

**1716,** Father Espinosa, Capitan Domingo Ramón, and the Frenchman Louis Juchereau de St. Denis stopped at the springs.

**1718,** Governor Martín de Alarcón crossed the river and named it the "Río de Inocentes."

**1721,** the Marqués de San Miguel de Aguayo visited the river.

**1727,** Brigadier Pedro de Rivera visited the river.

**1740s,** local Indian tribes encountered Lipan Apaches moving in from the north.

**1755,** missions from the San Gabriel River moved to the San Marcos, but the location is unknown. A large number of Apaches joined the mission.

**1756,** missions moved from the San Marcos River to the San Saba River, near present-day Menard, Texas.

**1758,** Comanches and other tribes destroyed the San Saba mission and began to dominate the region.

**1779,** Athanase de Mézières visited the springs and wrote of their beauty.

*From the Deep Past at the Springs*

would extend their power from Louisiana. But in their various efforts to save souls and defend the territory they claimed from the French, the Spanish visited the San Marcos Springs. They ultimately built missions and a settlement on the river and in so doing gave the river and the modern community their names.

Early Spanish maps and chronicles of expeditions mention the "San Marcos River," but they refer to the river we now call the Colorado. Accompanying the expedition led by Domingo Terán de los Ríos, the Franciscan Father Damián Massanet wrote on June 26, 1691, "We stopped on the banks of the San Marcos, which the French called the Colorado River because the soil was red and even the water seemed to be."

It is not always clear which river the Spanish described. However, we are sure the Terán de los Ríos expedition camped at the San Marcos Springs on June 20–25, 1691. While at the springs 110 of the Spaniards' horses stampeded, and the expedition recovered only 35 after spending several days searching for them.

Other Spaniards visited the springs, but none stayed for long. Governor Gregorio de Salinas Varona was there on June 27, 1693, with twenty soldiers and a pack train of ninety-six mules loaded with provisions to provide relief to the East Texas mission. His route, including the stop at the springs, became part of El Camino Real.

Sixteen years passed before the next recorded Spanish visit to the springs. In 1709 a small expedition set out to respond to a request from the Tejas people for a mission. Captain Pedro de Aguirre was the military leader, with fifteen soldiers. Father Isidro Félix Espinosa was the religious leader and was accompanied by Father Antonio Olivares. They arrived at San Marcos Springs on April 15, 1709. Espinosa wrote the following description of the region:

> Though it may seem a digression, Your Excellency, I cannot fail to mention in passing, that in addition to the fertility of the country exhibited by the variety of flowers, trees, and wild fruits, an abundance of hemp was noticed in the depressions of the ravines. This was so flourishing that it seemed to be cultivated, though it had received no other care than that of the liberal hand (of nature) that beautifies everything. The hemp found in the fields could supply all the wants of the Indian women. Besides this, the land seems to be suited to the cultivation of vines, a great variety of which are found growing wild on the hills. The vines are very large and resemble those of Castile. The bunches are larger and the grapes thicker, the skin being tougher; but the fruit is sweet and palatable. Mulberry trees are found everywhere along the arroyos and rivers. Their fruit is very sweet and the leaves large, as large as those of trees planted in orchards. The nuts are so abundant that throughout the land the natives gather them, using them for food the greater part of the year.
>
> In some rivers medlar trees are found like those of Spain. The variety of birds of various colors and sweet song is great. The deer and fawn are so numerous that they resemble flocks of goats and are met at every step. The buffalo is a singular beast among the wild life of this country.... Its spirit is weak and its courage short because on feeling itself wounded, even though it be in part not necessarily vital, it soon stops short and after a short while falls dispirited to the ground, blood issuing from its mouth. Its flesh is like that of the cows of Castile although better in taste and lightness. It constitutes the most common food of the nations that live in the neighborhood of the Tejas Indians, and of those in the hills where there is an abundance of buffalo. There are, besides, flocks of wild turkeys which are found at every step. There are bears, lions, tigers, foxes, and a great variety of other wild animals. Fish are so plentiful that there is not a creek, river, or pond where mullets, haddock, bagre, sea brim or buffalo fish, moharra, and every other species known are not found.

Father Espinosa returned to the springs seven years later with Captain

*The dependable water flow from the springs supported profuse plant and animal life for early residents.*

*In the springs region, deer are still so numerous that they resemble flocks of goats, as Father Isidro Félix Espinosa described in 1709.*

Domingo Ramón and the Frenchman Louis Juchereau de St. Denis. They visited the springs on May 20, 1716. St. Denis was one of the French traders who made the Spanish officials worry that the French rapport with natives would undermine the shaky Spanish control over the region.

For many years St. Denis traded contraband. He claimed to want to become a Spanish subject, but some Spaniards thought he was a French spy. Whatever the case, St. Denis was an expert on the geography of the region and undoubtedly shared his knowledge of the springs at the base of the escarpment.

The Spanish continued to visit the upper San Marcos because it was on or near El Camino Real, the King's Highway to their East Texas missions. Governor Martín de Alarcón crossed the river twice in May 1718 and named it the "Río de Inocentes." The Marqués de San Miguel de Aguayo was there on May 19, 1721, and Brigadier Pedro de Rivera on August 21, 1727. The Spanish made no effort to settle near the springs or the river because that was not their strategy, though in 1755 the springs and the river were briefly the site of a Spanish mission and presidio.

Ten years earlier a group of Yojuane, Deadose, Mayeye, and Ervipiame Indians had visited the San Antonio missions and asked the Spanish missionaries to establish a mission for them on the San Gabriel River, known then as the San Xavier. With a great deal of controversy the crown approved the project, partially with the expectation that the missions would prevent the hostile Lipan Apaches from moving farther south into Spanish territory and that they would discourage the French from trading with the natives.

The Franciscans from Querétaro established three new missions near present-day Rockdale in 1747. Presidios at Los Adaes, San Antonio, and La Bahia refused to contribute soldiers for their defense, so in 1750 the crown authorized a presidio for the missions. Captain Felipe de Rábago y Terán and fifty soldiers arrived in December 1751 to build the presidio. Conditions quickly deteriorated as the captain took liberties with the women of the settlement and violated the sanctuary of one of the missions. The missionary chaplain excommunicated the entire garrison. A priest and a soldier were killed, and the captain was implicated. The native people left.

Because of the slowness of horseback communication and the machinations of the conflicting church and government bureaucrats, nothing happened until almost four years later, when in August 1755, the missions and presidio were moved to the San Marcos River. Relationships with the Apaches had improved, and they had asked for missions to serve them. Perhaps the Apaches knew that the San Marcos location was desirable because it was a long way from their enemies, the Comanches. However, the next year the mission and presidio were moved to traditional Apache territory on the San Saba River near present-day Menard and given the name Santa Cruz de San Saba. In 1758 a combined force of about two thousand Comanches, Tejas, Tonkawa, and Bidai people attacked the mission, killed eight people, and burned the buildings. Santa Cruz de San Saba was the only Spanish mission in Texas destroyed by natives.

Some people speculate that the San Marcos mission was located above the springs. The owners of the mid-twentieth-century amusement park located there built a rock structure on the steep slope west of the spring lake that was reminiscent of a Spanish mission. One writer in the 1960s even speculated that the "mission" at the amusement park was built from the stones of the original bell tower, stones that were moved from the San Gabriel River. However, archaeological investigations at the probable site of the missions on the San Gabriel River found evidence of jacal walls—structures made of sticks rather than rocks. Captain Pedro Rábago y Terán reported in 1754 that soldiers at the presidio on the San Gabriel lived in thatched huts. Thus, the mission on the San Gabriel was probably not at all like the stone structures we see in San Antonio, nor was the temporary mission on the San Marcos River.

Not only was the mission probably not a stone structure with a large bell tower, it probably was not located at San Marcos Springs. It may have been just downstream from the springs, near the current sports auditorium of Texas State University–San Marcos, or it may have been at the site downstream that later became the location of the ill-fated San Marcos de Neve settlement in 1808, although no archaeological evidence of earlier settlement has been found there.

The Spanish wrote little about the San Marcos Springs until Athanase de Mézières' lyrical description. Mézières was a French nobleman, born in Paris, who became an agent of the Spanish government. He was assigned to Natchitoches in the 1740s, where he married St. Denis's daughter with the lovely name of Marie Petronille Feliciane. She was the granddaughter of the Spaniard Domingo Ramón, who led the expedition that founded the first mission at Los Adaes in East Texas. Sadly, the young woman with the pretty name died in childbirth a year after her marriage.

Many years later De Mézières described the springs on September 25, 1779:

> Having halted near the head of the San Marcos River, a worthy rival of the San Xavier (Brushy Creek) in respect to the conveniences which it offers for settlement, I have seen with wonder that it owes its origin to a huge rocky bluff, which emits from an ill-proportioned mount such a volume of water that it once becomes a river. One sees in the neighborhood several caves, with wonderful formations; here are some steps, an altar, frontal candlesticks, and a font; there curtains, festoons, flowers, images, and niches, all so clean that they appear to be in some one's charge. And there is no lack of benches, which invite the spectator to contemplate at leisure figures, some sacred, some profane, upon which nature has spent so much care that our Europe may well grieve at not being endowed with their equal.

De Mézières wrote this just before he died at age sixty. He had recently received a serious head injury when his horse threw him while he was riding between Los Adaes and Nacogdoches. Perhaps he drew comfort from relating the beautiful springs to the Parisian churches of his childhood.

De Mézières wrote that the valley of the San Gabriel River could be

*From the Deep Past at the Springs*

*The springs continue to flow from the "wonderful formations" that Athanase de Mézières described in 1779.*

irrigated, producing rich farms, with rock houses that would last from generation to generation. He surely recognized the same for the Colorado, San Marcos, and Guadalupe valleys. The Spanish were obviously aware of the economic potential of Central Texas, so why didn't they build their irrigation works and their farms, barns, and rock houses? Actually, they tried, as we will see later, just a little downstream on the San Marcos, but their efforts were unsuccessful. The primary reason the Spanish did almost nothing with the springs at San Marcos and the river was their complex and ultimately unsuccessful relationship with the native people.

We know almost nothing about natives that used the springs during the Spanish period. Terán de los Ríos noted meeting sixty Cantona people at the springs in 1691. Somewhere near the San Marcos River, or perhaps the Guadalupe, he encountered what he estimated to be two to three thousand natives of several tribes. In addition to the Cantona, other groups probably lived at the springs or used them. Those groups included the Muruam, Payaya, Sana, and Yojuane. Hunting groups from South and West Texas visited the springs, including the Catqueza, Caynaaya, Chalome, Cibolo, and Jumano. The Tonkawas, Apaches, and Comanches were latecomers but soon dominated the region.

Sixty-nine Tonkawa people accompanied de Mézières as he stopped at the springs on his way to San Antonio. He described how their lifeways were suitable for the country at the time:

> Their offensive weapons are firearms, bows, and spears; their defensive armament, skins, shields, and leather helmets with horns and gaudy feathers. The country being dangerous, through its being frequented by the Apaches, they use precaution; they explore the land, choose the most advantageous places to pitch camp, and post sentinels; they are exhorted morning and evening that their sleep be short and light; they arise at dawn to bathe; they give no chance by straying off from their march for being surprised by the enemy.
>
> The extreme neglect of the Indians to carry supplies would be criticized by one who did not know of their activity and sagacity in providing themselves with necessities. . . . In truth, one cannot exaggerate the inestimable benefits for which these natives are indebted to divine providence. The buffalo alone, besides its flesh, which takes first place among healthful and savory meats, supplies them liberally with whatever they desire in the way of conveniences. The brains they use to soften skins; the horns for spoons and drinking vessels; the shoulder bones to dig and to clear off the land; the tendons for thread and for bowstrings; the hoof, as glue for arrows; from the mane they make ropes and girths; from the wool, garters, belts, and various ornaments. The skin furnishes harness, lassos, shields, tents, shirts, leggings, shoes, and blankets for protection against the cold—truly valuable treasures, easily acquired, quietly possessed, and lightly missed, which liberally supply an infinite number of people, whom we consider poverty-stricken, with an excess of those necessities which perpetuate our struggles, anxieties, and discords.

De Mézières' Tonkawa companions were concerned about the Apaches, but the Spanish were concerned about both the Apaches and the Comanches. All three tribes were newcomers to Texas. The Apaches, like the Navajos to whom they are related, had migrated into New Mexico and Arizona about four hundred to eight hundred years earlier, probably from Canada. They pushed the Tonkawa from the northern buffalo ranges during the 1600s.

While the previous several hundred years may have been relatively stable, life for Indians in the 1700s was in major turmoil, largely the result of the horse, which the Spanish had reintroduced to the New World. For the hunter-gatherer people the horse was revolutionary. The Indians could more easily follow and kill the buffalo—and they could roam and raid afar. Thus, successive waves of people on horses moved into the springs area from the north and west—Tonkawa, Apache, and Comanche.

The Apaches quickly learned to use escaped Spanish horses and

*Although people have come here for thousands of years, the springs continue to be a refuge for nature.*

became feared raiders on Spanish and Pueblo settlements. Like the Navajo, they called themselves the *Diné,* meaning the people. However, the Zuñí called them the *apachu,* meaning "enemy." But like the Navajo, the Apaches were farmers as well as raiders. They were thus exposed to raids by the totally footloose Comanches, who pushed the Apaches into Central Texas.

The Comanches had been mountain hunter-gatherers, part of the Shoshone people. After they got horses they became the most feared people of the region. They too called themselves "the people," but the Utes called them *Komántcia,* which meant "anyone who wants to fight me all the time." They raided and traded the booty of their raids. The Spanish first encountered them in Texas when a Comanche scouting party came to San Antonio in 1743, looking for Apaches, whom they assumed were allies of the Spanish. By the late 1750s, Comanches dominated the region north of San Antonio, including San Marcos Springs.

The people who stopped at the springs were not simply roving hunters and raiders. They were also traders. They were interested in the Spanish for three reasons. First, the Spanish livestock and possessions were easy prey. Second, the Spanish in Texas gave gifts to maintain the support of cooperative tribes. Third, the Spanish were powerful allies against enemy tribes. There was a constant complex intrigue of shifting promises and allegiances between the Spanish and the various tribes, not only in Texas, but in New Mexico and south of the Rio Grande as well. So life for people who visited the springs was most uncertain.

Movies, and even some historians, have left us with a view of the Spanish as powerful conquistadors interested only in gold and souls and a view of the Indians, especially the Comanches, as desperate and cruel people. As with most stereotypes, both of these are somewhat true, but they miss the full complexity of the relationships between the various peoples who stopped at the springs. These complex relationships may explain why no one, including Indians, made a permanent home for themselves at the bountiful springs during the early days.

Although the Spanish were motivated by gold and souls, they were officially governed by the Laws of the Indies, specific policies and regulations regarding treatment of native people and protection of their rights under the law of the king. In Texas the overwhelming concern of the Spanish was to keep out the French, so the Indians were essential allies. By policy, if a tribe would swear allegiance to the Spanish king and would not kill or steal from the Spanish, the Spanish would accept them as brothers and as children of the king. However, this policy was often not the practice before the 1760s.

In 1785 the Spanish made a treaty with the Comanches and other tribes in Texas. Spanish Texas slowly developed for a period of about twenty-five years. However, the Mexican War of Independence began in 1810, distracting Spain from this northern edge of their empire. Except for events far removed from the San Marcos Springs, the Spanish may very well have developed the resources they recognized there.

Indian legends say that the springs were a gathering place, a place of refuge, where enemies would come together in peace. We don't know, but we do know that no people took the springs as their home until Anglo Americans and their black slaves showed up in the 1840s with a very different perspective on the land and the water.

# CHAPTER 4

# Anglo Americans at the Springs

SPANIARDS SOUGHT GOLD and souls. There was no gold on the San Marcos, and the souls had their own ideas of salvation, so the Spanish left the river. But the aggressive bunch of bear shooters, tree cutters, sod busters, and dam builders to the northeast were extremely interested in rich black land alongside a constant flowing spring-fed river.

American settlers in Green DeWitt's colony, including my own ancestors, congregated around Gonzales at the confluence of the San Marcos and Guadalupe rivers in 1825. Gonzales played a major role in the Texas Revolution and was abandoned and burned. But the Texian army led by Sam Houston won, established a republic, and set out seriously to Americanize the land, including the San Marcos River.

The city of Austin, capital of the new Republic of Texas, was beyond the frontier and not easily accessible. The old Spanish road, El Camino Real, went from San Marcos northeast to Bastrop. No road connected Austin to San Antonio, the state's largest city. In late 1840 the Republic of Texas established Post San Marcos at the San Marcos Springs and built a road from San Marcos to Austin, which is today's Post Road and Old Stagecoach Road wending west of Interstate 35. On October 21, 1840, Adjutant and Inspector General Hugh McLeod wrote a letter to the

*The San Marcos Springs reflect more than 150 years of active use and change. Most of the facilities of the Aquarena Springs resort will be removed, and the site will be restored to a more natural condition. The water covered with vegetation in the background is the "Slough," where Sink Creek enters Spring Lake.*

president of the Republic of Texas, General Mirabeau Buonaparte Lamar, describing the springs area and progress on the road:

> The Company under Capt. Weihl have done themselves great credit in cutting a road of 3½ miles thro' a dense bottom interlaced with brier, grapevine—(one in view now measuring eleven inches through) and under growth of the most stubborn kinds— The term bottom however does not properly apply, for the spring or springs for there are several being the extreme head of the river, it never rises nor falls, nor are there annoying insects—The springs rise along the foot of a mountain, whose highest peak (it is irregular) is about 800, to 1000 feet above the grand prairie which it overlooks—Fairy Land cannot excel it in the beauty of its landscape, nor will the highlands of the Hudson compare with the bold, yet softened scenery of its mountain views—Towering

hills arise on every side, but the bleak baldness that would chill the blood in a norther clime, is vieled here, in the perpetual verdure of the live oak—

The fort was occupied only from October 1840 until March 1841, but this location at the junction of the old road to Bastrop and the new road to Austin was ripe for a town to serve travelers. In contrast with today's rapid land development in the region, it was not until 1846 that the town was surveyed. Although the river was attractive, the town was located to the west of it to avoid floods.

In 1846 William A. McClintock, an early traveler in Texas, described

*There was no lake when Anglo Americans first came to the springs, but the wetland formed by the springs was probably foggy on a winter's day.*

Anglo Americans at the Springs

the springs on what he thought was the Blanco River:

> Two miles north of St. Marks we crossed the Blanco, a mountain torent of purest water, narrow and deep, there is the finest spring or springs (for they are not less than 50 in a distance of 200 yds.) I ever beheld. These springs gush from the foot of a high cliff and boil up as from a well in the middle of the channel. One of these, the first you see in going up the stream, is near the center, the channel is here 40 yds. wide, the water 15 or 20 feet deep, yet so strong is the ebulition of the spring, that the water is thrown two or three feet above the surface of the stream. I am told that by approaching it in canoe, you may see down in the chasm from whence the water issues. Large stones are thrown up, as you've seen grains of sand in small springs, it is unaffected by the dryest season. I am persuaded that the quantity of water which is carried off by this stream in the course of a year is greater than that by the South Licking, it is about 60 feet wide and 3 feet deep on an average, with a curant of not less than ten or fifteen miles per hour. Great numbers of the finest fish; and occasionally an alligator may be seen sporting in its crystal waters.... In the eddies of the stream, water cresses and palmettoes grow to a gigantic size.

Colonists moving into Texas from the East must have been disappointed by most Texas rivers. Coming from the East, where rainfall is more constant and where forested watersheds help moderate runoff, they expected rivers to flow more or less regularly. Eastern rivers had seasonal floods and low flow, but they were dependable for transportation and waterpower. Those variations were much more extreme in most Texas rivers. The Brazos and Colorado would flood mightily in the spring and early fall and at odd times just to keep you guessing, but their flow might be just a trickle in the summer and winter.

Imagine Moses Austin and his son Stephen looking at Spanish maps of Texas as they planned their colony. Those men from Connecticut would have seen the Brazos and Colorado rivers as crucial transportation links to ensure the success of their colony. How disappointed they must have been when they realized that those rivers did not behave like the Connecticut, the Hudson, the Mississippi, and the Missouri. If anything, the rivers were a pain—they were not very useful, and they flooded mightily.

But the San Marcos and other rivers fed by springs from the Edwards Plateau behaved like rivers should. Their flow was relatively constant and provided a dependable source of water. These rivers were not big enough for transportation, but they were valuable sources of waterpower for grain mills and sawmills.

In June 1846 the German geologist Dr. Ferdinand Roemer described coming upon the San Marcos River as he traveled from Austin to New Braunfels:

> The course of the San Marcos, the only important river between the Colorado and the Guadalupe was indicated from a distance by a narrow strip of forest a little beyond the other side of the Live Oak Springs. Soon after we descended into the beautiful fertile valley. Before arriving at the San Marcos we had to cross the Rio Blanco, a pretty, clear stream.... A bottom, three miles wide, covered partly with brush, partly with luxuriant trees, separates it from the San Marcos. The latter is a beautiful river abounding in water, which flows rapidly and is of such magical clearness as can be found only in the rivers of western Texas. The springs of the San Marcos are only several hundred feet above the ford. Surrounded by the evergreen bushes of the palmetto (*Sabal minor Pers.*) and shaded by stately forest trees, to whose pinnacle the mighty grapevines climb, resembling anchor ropes, they break forth under the thick limestone boulders with such tempestuousness and volume of water that they could turn mills at their immediate source.

Roemer also observed the effects of floods on the river. "Unfortunately the broad fertile bottom of the San Marcos will never be suitable for agriculture since it is subject to inundation," he wrote. "While riding through it we noticed with astonishment dry cane

*The source of the river is visible through the clear water and the glass bottom of one of the tour boats.*

and limbs hanging fifteen to twenty feet high above the ground from the trees, which the spring floods had left there."

But about a month later Roemer was again at the San Marcos and seemed to have a different impression. "Our first halt for the night was made at the San Marcos, since we had left New Braunfels in the afternoon. Everything appeared more animated at this place than on a recent visit, for a company of mounted rangers was stationed here, and several families with large wagons had arrived to found a new settlement," he wrote.

*General Burleson's primitive dam remains in place after more than 150 years of repairs.*

"In fact, a more advantageous and pleasant place for a settlement could not be imagined than this parklike little prairie, surrounded on one side by the forest fringing the San Marcos and on the other by the steep hills, the beginning of the higher hill country."

The families Roemer saw included the Moons, Merrimans, Lindseys, and Burlesons. Although General Edward Burleson was only one of the original partners in the new town, he more than any other seized on the opportunities the San Marcos River presented. Burleson had developed a mill on the Colorado River at Bastrop in 1830. After moving his family to San Marcos, in 1849 he built a dam just below the springs at the head of the San Marcos River.

Burleson's dam drastically changed the headwater springs from a spring-fed marsh with geysers to a deep clear lake with very utilitarian purposes. The lake was not the objective in 1849, but the "head" of the water—the height to which it was elevated above the original river surface level—was of paramount importance. This head of 8 to 12 feet provided power in those times before internal combustion engines and electricity. At the time of Burleson's death in 1851, this dam provided a fall of water that turned waterwheels to power a gristmill, sawmill, cotton gin, and cotton press. These industries, located

*Burleson's old raceway now flows beside a restaurant, and the dam continues to impound water for Spring Lake.*

on the road between San Antonio and the upstart state capital, provided the kernel of what became the city of San Marcos.

The issue of how the headwaters should be developed and used was a controversial one, as it has been for years. In 1876 the state of Texas proposed to locate a penitentiary at the headwaters. Isaac Julian, editor of the *Free Press,* expressed his strong opposition to the proposal:

> Do we want to see that glorious SPRING, by which, in seasons of drought, men and animals are driven to resort for 10 miles around, thus defiled, and the grand flood of LIVING WATER flowing therefrom, more precious than gold to this thirsty land, become the sewer to this horrible concentration of filth and poison, taking its slimy way alongside our beautiful town, leaving defilement, disease and death in its course? . . . We feel sure that the almost universal response in this community would be one emphatic 'No!' Let it then be given, of necessary, and may it be heeded by the powers that be.

Before we move downstream to San Marcos, we should continue to move through time at the headwaters, because one of Texas' earliest and most famous resorts developed on the incredibly clear waters held back by General Burleson's dam.

## A. B. and Paul Rogers

The Rogers family was the epitome of early twentieth-century optimism and entrepreneurship. A. B. Rogers's family moved from Ellis County, Texas, to San Marcos in 1873, when A. B. was two years old. He started his business career as a clerk in a hardware store, but he soon established a furniture store and funeral parlor. W. A. Pennington bought the funeral home in 1944, and his family continues to operate it. In 1911 Rogers established a truck garden on the San Marcos River and then converted it to Rogers Park, the site of the current Rio Vista Park. Rogers also developed Wonder Cave and owned three ranches, but his lasting mark on the San Marcos River began in 1926. The *San Marcos Record,* on March 12, 1926, reported:

A. B. Rogers, owner of the famous Rogers Park and Wonder Cave, has purchased 125 acres of land from the San Marcos Utilities Company including the head of the river, and will make of this one of the great playgrounds of Texas and the Southwest.

The tract includes a spot of historic interest in the site of the old home of Gen. Edward Burleson, the home built around the year of 1847. His two-room log house built on the summit of the hill overlooking the beautiful San Marcos river at its source, remained until a few short years ago, when it fell to the ground, leaving only the hearthstones and the rock foundation to tell the story. It was on this same hill, just a few hundred yards back of General Burleson's home, that the first school house of San Marcos was built in 1849, but it lay too remote for safety and was abandoned for one built in the heart

*General Burleson built his homestead above the San Marcos Springs. Photo reproduced from the San Marcos-Hays County Collection at the San Marcos Public Library.*

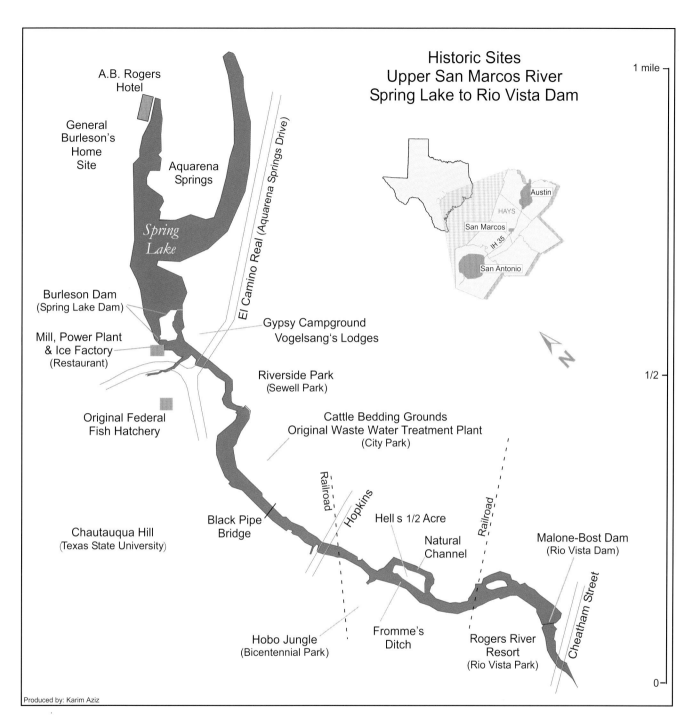

of the little town and used for several purposes of school house, court house and church.

The acreage that stretches from the east side of the river to the Post Road and [sic] will be given over to golf course.

A big motor boat is now being built to clear the growth out of the headwaters of the river and eventually this part of the river and its banks are to be made into a delightful bathing pool and beach. A walk is to extend along the banks to the head of the river, built for the convenience of

# For Sale—San Marcos Springs

The beauty of the headwater springs was expressed as real estate value as early as 1877, when an advertisement in the September 13 edition of the *Galveston Daily News* stated:

SAN MARCOS SPRINGS
PLACE FOR SALE $5500
THIS VALUABLE PROPERTY,
adjoining the town of San Marcos, comprises 345 ACRES.
It encloses the entire source of the beautiful San Marcos River
70 ACRES IS CHOICE

pedestrians, while roadways will be built making a scenic drive for motorists. On the plateau that rises sheer and precipitous from the river's bank near the head, Mr. Rogers proposes to build his own home and a number of modern residences. Campsites are to be duly fitted up. It is planned to make the place inviting to the traveler and to the pleasure-seeker and equally to the stay-at-home vacationist.

Almost exactly three years later, on April 22, 1929, Rogers opened his Spring Lake Park Hotel overlooking

*Burleson's dam formed a lake in the upper part of the San Marcos, now called Spring Lake. Postcard is dated 1907. From the postcard collection of Jim Pape.*

*The device suspended over the edge of this steam-powered stern wheel work barge may be a dredge for clearing aquatic vegetation. Photo reproduced from the San Marcos-Hays County Collection at the San Marcos Public Library.*

was polluted by the city's old wastewater treatment plant. But surely the young men enjoyed Spring Lake. Certainly, Preston Connally did.

Preston, a San Marcos native, was born in 1920. In his late teens he made a diving helmet from a water heater tank, complete with a porthole, padded shoulder cutouts, and an air hose that led to a pump on the surface. Preston would don the helmet and walk on the bottom of Spring Lake. Air in the helmet would hold the water at about chin level if he was skillful in the descent. Doris, Preston's wife, said she didn't do it right, and water came up into the helmet, marking the end of her diving career, fortunately without disaster.

The end of WWII brought new affluence and a desire to enjoy life that had not been possible through the

the headwaters, a gala event marked by a golf tournament, a band concert, dinner, and a rooftop dance. Sadly, 1929 was a bad time to open a resort hotel. The hotel struggled for a few years in the early part of the Depression. Locals continued to enjoy rooftop dances until an argument resulted in a murder there. Rogers leased the hotel to a hospital and then to the Brown Schools, a school for troubled children, as a treatment facility.

The 1930s and the wartime were a quiet period for the headwaters, even though San Marcos hosted Gary Field, a training site for Army Air Corps navigators. The base commander declared the San Marcos River off-limits, because the water

*A. B. Rogers opened his Spring Lake Park Hotel in 1929 at the head of the San Marcos River. The hotel is now part of the Texas Rivers Center. From the postcard collection of Jim Pape.*

*A. B. Rogers's 1929 hotel,* above, *now houses offices for the Texas Rivers Center. The old resort facilities,* bottom, *have hosted thousands of schoolchildren who learn about aquatic life and the Edwards Aquifer.*

Depression and the war. The Rogers family responded in a big way. In 1946 A. B. Rogers's son, Paul, built a fishing pier. He also put a glass viewing port in the bottom of a wooden boat with a canvas shade over the top. This first glass bottom boat on the San Marcos River was very popular, so Rogers went to Florida to see the boats used there. He modified the Florida boat design and contracted with Jack Warner of San Marcos to build five boats.

A *Dallas News* writer in 1949 observed that Springlake Marine Gardens, Silver Springs, Florida, and Catalina Island, California, were the only places in the United States where visitors could enjoy glass bottom boats. That same year Paul Rogers took his wife and children to Weeki Wachee Springs in Florida. While there he recognized that the clear water of Spring Lake could host an underwater theater and started planning to build one, including dredging the east side of Spring Lake to provide sufficient depth. He formed a partnership with Marine Studios of Marineland, Florida, and completed the underwater theater, opening it in September 1950, naming his attraction Aquarena Springs.

Although Rogers bought out his partners in 1951, Marine Studios brought important experience to the project through its president, W. Douglas Burden. Burden had been associated with the American Museum of Natural History since 1923 and was

*The east bank of Spring Lake had not yet been excavated for the submarine theater in this photo from the early 1950s. From the collection of Jerry and Jim Kimmel.*

a trustee of the New York Zoological Society. He led expeditions to Asia and produced documentary films.

The submarine theater at Aquarena Springs received national recognition. An illustrated article in the June 1952 issue of *Popular Mechanics* described this unusual feature:

At San Marcos, Texas now boasts a venture unique to both the entertainment and educational worlds—a theater which allows an amazed audience to witness an hour long program beneath the surface of a crystal-clear lake. It is a submarine theater which, when a special ballast tank is flooded, takes its cargo of people below the surface.

The arena holds three million gallons of water and is 30 feet deep. Between shows, a pump spills 10,000 gallons of fresh water into the arena per minute for a constant supply of clear water. The water is taken directly from an encased spring.

Submerging 42 inches in 11 minutes, the theater will accommodate 100 spectators, who witness the underwater routines through plate-glass windows. The submarine's entrance, which is above water even when the theater is submerged, remains open at all times.

Some 50 tons of steel and 20 tons of concrete went into the submarine theater's structure. It requires 15,500 gallons of water for submersion.

With inspiration and advice from Florida, Rogers apparently envisioned a "marine garden" as it was called in the 1949 *Dallas News* article. With this marine vision, the grand opening of Aquarena featured trained California sea lions along with the young Aquamaids. But "marine" implies

*The original glass bottom boats will continue to delight visitors as they have for more than fifty years.*

the ocean, and sea lions are saltwater animals, not freshwater. Although Aquarena employed a professional to care for the sea lions and kept them in a saltwater pool, they became ill, and the show dropped its inappropriate marine emphasis.

Low, sprawling limestone buildings on the east side of Spring Lake formed the entrance to Aquarena and provided space for a restaurant, souvenir shop, and administration offices. These buildings, now scheduled for demolition, were carefully designed and used modern construction methods of the day. The exteriors were partially Cabin Creek ledgestone and partially a concrete and asbestos composite molded under pressure into a corrugated surface. Exterior trim was California redwood, and the interior was paneled with heart-grain plywood. The architect, George Walling of Austin,

*Texans rarely had the opportunity to peer into deep clear water. Photo reproduced from the San Marcos-Hays County Collection at the San Marcos Public Library.*

dealt with the high water table at the site by drilling large diameter holes through the alluvial material down to hard clay and then filling the holes with concrete.

Paul Rogers and his managers thoroughly understood the fickleness of their audience. People constantly want new things, and Aquarena provided them. In 1955 Rogers doubled the size of the main building, renovated the restaurant, added forty seats to the submarine theater, and expanded the historic-themed Texana Village. Aquarena was expanding just as Disneyland was opening—who inspired whom?

Somewhere along the way in the late 1950s a swimming pig joined the act, first named Magnolia, then Ralph. The succession of young pigs named Ralph were obviously heartier than sea lions.

From 1961 to 1963 Rogers spent $300,000 for additions and improvements. He commissioned the nationally recognized artist from Wimberley, Buck Winn, to design a sculptural entrance to the park and fountains for the lake. The complex fountain used electronic controls to create a water curtain eighty feet long, with periodic streams of water shooting thirty feet into the air. Winn said, "We wanted to get away from the stereotyped fountains with bronze figures of women pouring water out of a jar or frogs spitting water."

The early 1960s additions also included a tropical garden on the steep west side of the lake and an aviary behind the hotel. Visitors could board the new sky ride on the east side of the lake and ride gondolas to the hanging gardens. The sky ride, built by Von Roll Iron Works in Berne, Switzerland, could carry three hundred passengers per hour.

Tarzan, in the person of Johnny Weissmuller, visited Aquarena in March 1965 to film the swimmers for a television series. He screamed his Tarzan call in the restaurant and then told a San Antonio reporter that he "learned it as a kid, entering yodeling contests around Chicago."

Only a few months after hosting Tarzan, Paul Rogers died an untimely death at age sixty-seven. He left plans for a new $1 million motor hotel to be built around the headwaters. The motor hotel was never built, but the Rogers family maintained ownership, and Gene Phillips provided leadership to continue Rogers's momentum for Aquarena. A volcano was added to the underwater stage in 1967. Scott McGehee became general manager, adding the gyro tower in 1978, replacing the submarine theater after the 1970 flood, and renovating the old icehouse building at the dam into a restaurant in 1982.

John Baugh purchased Aquarena from the Rogers family in 1985. He made few modifications, but the whole context had changed. Aquarena was no longer unique. Six Flags Over Texas had parks in two major Texas cities. SeaWorld was opening

*Even politics were underwater at Aquarena Springs. Photo reproduced from the San Marcos-Hays County Collection at the San Marcos Public Library.*

to find, especially with attractive young performers swimming underwater ballet. My little sister and I, as many other young Texans did, rode in our parent's unair-conditioned Ford station wagon from a hot dusty little town to the tropical lushness of the headwaters of the San Marcos. Twenty years later we took our own children, and then they took theirs. But things change. In 1994 Texas State University–San Marcos bought Aquarena for $7 million dollars and set out to find a new direction for the magnificent headwaters.

Robin Ward, who worked as an "underwater performer" at Aquarena Springs when she was a college student, was part of Aquarena's transition from amusement park to nature center. When asked in 2001 to describe her experiences there, she wrote:

> Once I descended beneath the surface of Spring Lake my life was forever changed. I was hired as a Mermaid in the underwater show, The Legend of Singing Wolf. On any given day at work I spent anywhere from three to five hours under water with an air hose as my only connection to terrestrial life. During most shows, more than 200 people sat within the Submarine Theater watching the performers feed the aquatic animals, coax "Ralph" to swim laps, and hold their breath while performing underwater ballet. I felt exhilaration every moment that I was beneath the water performing in front of hundreds of tourists.

in San Antonio, and many Texans had the money to fly to Disney World, Disneyland, Cancun, and Cozumel. Growing sensitivity to environmental issues and the Endangered Species Act raised questions about Aquarena's impacts on the sensitive headwaters ecosystem. Ralph the Swimming Pig was no longer relevant and maybe not even appropriate.

Aquarena was unique in Texas, where naturally clear water is hard

But it was the energy of the water from the springs that cast a spell on me....

As the University began the evolution of Aquarena Springs, the theme park, to Aquarena Center, the educational nature center, I felt drawn to return to the headwaters of the springs. I became involved in the care of the native endangered species that are on exhibit in Aquarena's aquaria. I obtained a permit to maintain endangered species populations in captivity through the State and Federal Fish and Wildlife Agencies. I have since had the fortunate opportunity to witness the breeding between a pair of endangered Texas Blind Salamanders, which rarely occurs in captivity. In the past four years I have coordinated environmental and ecological tours with local schools, given glass bottom boat tours of the underwater springs, and assisted in educating park employees.

I too have changed with the park. I traded my mermaid costume for a degree in Environmental Studies and Resource Management. I plan to seek certification through the National Association of Interpretation as an instructor, where I hope to have the opportunity to educate others in the importance of protecting our natural resources.

Jerry Supple, president of Texas State University–San Marcos, knew that the headwaters was a special place and deserved special treatment, but the university's purchase

*The wetland boardwalk in the Slough offers an intimate view of nature.*

was controversial. Townsfolk felt that they had lost an income-producing asset, and the university may not have fully recognized the difficulty of converting the old amusement park into some other kind of attraction that would support itself.

Under the management of Ron Coley, the renamed Aquarena Center has refocused itself toward environmental education and nature tourism. Ralph the pig was dismissed early on. The glass bottom boat tours provide high quality interpretive information about the headwaters ecosystem, its endangered species, archaeology, and the Edwards Aquifer. Thousands of schoolchildren visit Aquarena Center annually for environmental education programs that are coordinated with state curriculum requirements.

In 1997 Andrew Sansom, executive director of the Texas Parks and Wildlife Department, proposed to the university that the two agencies cooperate to convert Aquarena into the Texas Rivers Center, similar to the Texas Parks and Wildlife Lakes Center at Athens and its Sea Center at Lake Jackson. The pace set by this combination of state, federal, and university bureaucracy is, shall we say, deliberate. Renovations were completed in 2006 to use the hotel for university office space and an interpretive center. The university plans to remove the buildings on the east shore of the lake and restore the area with native landscaping, and ultimately to build an interpretive display at the current site of the swimming pool. Funding sources and potential income are uncertain, but in the meantime families continue to experience the crystal clear waters of the San Marcos springs.

CHAPTER 5

# San Marcos the River, San Marcos the Town

PROBABLY NO ONE had deeper experience with or stronger opinions about San Marcos the river and San Marcos the town than C. W. Wimberley, a local historian and writer. He was the grandson of Pleasant Wimberley, whose prolific family settled west of San Marcos at the junction of the Blanco River and Cypress Creek in 1874, now the village of Wimberley. C. W. Wimberley's father was in charge of grounds maintenance at Southwest Texas State Teachers College. Wimberley is important to this chapter of the river's story not only because of what he wrote about the river, but also because his boyhood on the river defined the boundaries of the San Marcos segment. His river included the headwaters, but he had the most to say about the segment from Burleson's dam to Thompson's Islands. This Texas Huckleberry Finn lived with and in this part of the river, and we will use his boundaries to relate the river and the city. We will work our way downstream from Burleson's dam to Thompson's Islands as we work our way through time.

Geographers say that some cities developed at a break in transportation—a place where goods were transferred from ships to wagons, for example. Because there was delay in making the transfer, there was need for lodging, food, and services, and that created business opportunities for permanent residents. El Camino

*The San Marcos River truly is the heart of the city of San Marcos, flowing through Texas State University and the city's extensive riverside parklands and providing a congenial setting for the Activity Center and the San Marcos Public Library.*

Real crossed the San Marcos River at several places in the vicinity of the modern city, including a ford near the current Aquarena Springs Drive bridge. Although this is not a classic break in transportation like a seaport, the ford and later the dam and mills established a point of business and social activity.

Billy (William W.) Moon, his wife Sophronia, and their small children settled at the present site of San Marcos in 1845, the first Anglo settlers in what would become Hays County. Caton Erhard opened a store and post office in 1847. By 1848 stage coaches ran from San Marcos to Austin and San Antonio, and the town also became a shipping terminal for agricultural products headed to the Texas coastal ports. San Marcos was designated as the county seat of Hays County in 1851. In that same year General Edward Burleson, William Lindsey, and Dr. Eli T. Merriman obtained 640 acres from the Juan Veramendi grant and laid out the town, with its square seemingly far enough from the river to avoid floods.

Burleson's dam, which powered a gristmill, sawmill, and cotton gin, contributed to the growth of San Marcos. Albert S. McGehee, a descendent of one of the early settlers in the area, described the original power machinery: "The power unit for the mills was built by craftsmen Greenberry Ezell and William Firebaugh. The large, turbine-type water wheel was hewn from walnut, the octagon-shaped shaft of which was 15 inches in diameter and 18 feet long. This shaft, still basically sound after 80 years, was dredged from the river in 1928, by the late Rufus Wimberley."

Burleson's dam was primitive but effective. C. W. Wimberley described its construction in detail.

*Burleson's dam became the site of the San Marcos electric power plant before this postcard was mailed in 1911. From the postcard collection of Jim Pape.*

> This old relic of the past is worth your study—to see big cedar log piling angled across the river basin, boarded with smaller heart cedar logs fronted by a rock-filled barrier above their height at the width of a country lane, then sloped to ground level, covered with heavy and light gravels to fill the holes between these rocks, topped with clay hauled from the clay bluff beside Spring Lake to seal the top and slope of the structure. And then cottonwood trees were planted along its crest with a growing root system to reinforce the old structure so that it could withstand the torrents of the floodwaters to come.

This is a far cry from our modern massive concrete dams or even from the masonry dams the Spanish built in the New World. The secret to Burleson's dam and other early dams on the San Marcos is that they were fairly easy and cheap to build, easily damaged, but easily repaired. Of Burleson's dam, the 1895 Hays County Irrigation Records Book states, "The time when the work was commenced on the construction of a dam across said river at or about the same place where the present dam is located was about the year 1849, and the same has been maintained by repairs and the construction of new dams from time to time continuously until the present time."

Burleson's dam has continued to be damaged by floods, the latest in 1998. Major repairs were done in 2000.

The spring-fed river's dependable flow provided power for an ice factory in 1883 and an electric generating plant on the east side of the river in 1884. That year Tom Code, the owner of the mill, built a bathhouse at the mill site that allowed women to enter the river through a hole in the floor, so that they would not be seen "bathing" in public. Code rented rowboats and advertised his place from Canada to Mexico. Tourism began early on the San Marcos.

The twenty-fourth Texas Legislature, in 1895, passed a law with an incredibly long title that expressed the attitude of the day toward rivers: An Act to Encourage Irrigation and to Provide for the Acquisition of the Right to the Use of Water, and for the Construction and Maintenance of Canals, Ditches, Flumes, Dams, Reservoirs, and Wells for Irrigation and for Mining and Milling and the Construction of Water Works for Cities and Towns and Stock Raising. This law required a description and map of all water diversions from rivers.

The Hays County Irrigation Records Book includes six dams or other diversions in the first 1 3/4 miles of the San Marcos River. It doesn't appear that the good folks of San Marcos needed much encouragement to use their river.

T. U. Taylor, the first dean of engineering at the University of Texas in Austin, described the Burleson dam in *The Water Powers of Texas*, published by the U.S. Geological Survey in 1904:

> The first dam on the San Marcos is about one-fourth mile from the head springs. The dam is about 400 feet long, the eastern section crossing the channel at a right angle for two-thirds of the length, the remaining portion deflecting parallel to the west bank to form a fore bay. The dam is constructed of earth and piling and has a maximum height of 15 feet, and develops a head that varies from 8 to 12 feet. The water power is generated by two 35-inch Leffel turbines of the Samson type, and one 48-inch Morgan Smith Success turbine. An auxiliary Atlas steam engine of 110 horsepower is used to supplement the water power. The plant belongs to the San Marcos Electric Light Company, the San Marcos Water Company, and the San Marcos Ice Company. They are distinct companies. The building is made up of a boiler room 32 by 36 feet, an electric light and pump room 40 by 60 feet, an ice room 30 by 70 feet, and a storage room 20 by 40 feet. The electric energy is generated by one Warren 90-kilowatt generator and one Westinghouse 60-kilowatt generator. The pump is a Gordon Maxwell 10 by 12 duplex, with a capacity of 500 gallons per minute, but there is soon to be installed one Fairbanks 10 by 12, with a capacity of 1,000,000 gallons per day.

The original ice factory produced 1 ton per day, but by 1909 the expanded and modernized plant produced 15 tons of ice and 400 gallons of ice cream daily. The four thousand residents of San Marcos apparently enjoyed their ice cream.

At least two people died in the mill's heavy machinery. A plant attendant fell into the turbine, which ground him to pieces. The son of another attendant, Ed Beidler, fell into the set of 3-inch continuous ropes that served as the pulley belt to transfer power from the turbine shaft to a huge flywheel.

Electricity generated at the dam powered an extensive irrigation operation above the headwater springs. In 1905 William Green applied for a permit to irrigate 975 acres by pumping water from the lake formed by Burleson's dam. The water flowed through three canals, of 2,000 feet, 5,500 feet, and 7,000 feet. This innovative and diversified agricultural enterprise, named Riverhead Farms Inc., was headquartered at the Edward Jon Burleson house on Lime Kiln Road. Much of the land surrounding the upper part of the river produced a great variety of vegetables and fruit plus hogs and cattle, which were shipped out by railroad. Green owned 1,667 acres in 1905 and ultimately farmed perhaps as much as 2,500 acres.

Other vegetable farmers used the river to irrigate much of the land from the headwaters to Thompson's Islands, about 2 miles downstream. Before refrigerated railroad cars were invented, the growers used ice from the icehouse at Burleson's dam to preserve the vegetables during shipping. They crushed the ice and loaded railroad cars with alternating layers of ice and vegetables. Residents said San Marcos smelled like an onion.

The lake also supplied water for the growing city of San Marcos. The San Marcos Water Company, owned by Eugene and Ed J. L. Green, piped water from the lake to residential customers.

The river's power was insufficient for the increasing demand for electricity in the early twentieth century. By 1927 the only product the river's flow helped produce was ice, but some time before that the managers had installed a diesel engine to run an auxiliary generator to supplement the river's power. The engine had six cylinders, each 2 feet in diameter. W. C. Wimberley said, "When this six-cylinder giant roared into action with the delicacy of a freight train passing through your living room to let everyone across town know where their electric power came from, you may have reconsidered the simple life when you could blow out the lamp, let the cat out, and get a good quiet night's sleep."

The ice plant continued to operate until 1976, producing 40 to 45 tons per day. The Aquarena Springs organization bought the ice plant building in 1977. In 1982 they opened Peppers at the Falls Restaurant in the renovated and expanded building. In 1997 Joe's Crab Shack, part of a national chain, replaced Peppers.

## Sessom Creek to the Old Fish Hatchery Building

This reach of the river, most directly adjacent to the town, tells a variety of stories, from escaped bears to famous water pageants. In 1845 McCulloch's Mounted Volunteers camped on the west side of the river, awaiting service in the U.S. war with Mexico. Caton Erhard built a store to serve the soldiers, but they left after a few months, and he moved his store closer to the new San Marcos square. The road from San Antonio to Austin forded the river just below the dam. Gypsies and other traveling people camped in the area on the east side of the river. C. W. Wimberley wrote of Gypsies' bears that escaped there:

> In my time, the travelers' campgrounds across the river between the dam and the highway were known to some as the Gypsy Campgrounds where they, with their horse-drawn

*Workers process spinach to load on a train car. Much of the land surrounding the upper river was irrigated farmland. Photo reproduced from the San Marcos-Hays County Collection at the San Marcos Public Library.*

> vans and travelers of similar stripe, spent days or weeks beneath the tall willows beside the river.
>
> I was on hand when the trainer emerged from one of the tents across the river leading five bears by their leashes. On reaching the river, he released his charges to frolic and splash about in the water. Tiring of their sport, they were lying about in the water when the driver of a buggy coaxed his horse into the crossing from my side of the river, and the show was on.
>
> Sensing the bear's presence, the horse bolted, splashing his way upstream, exiting the river to cross the back lawn of the ice factory, headed yonder.
>
> In their turn the bears panicked

> from the water to take refuge in the treetops of the highest willows in the campground, and from my ringside seat the show continued.
>
> Ignoring their trainer below, the bears spent the night clinging to their high perches while he spent the night below beside a small fire and I spent the night sneaking in and out of my bed to keep my ring-side seat warm.
>
> At dawn the show was over. Clambering down to have their breakfast, the bears submitted to having their leashes again attached to their collars.

By the early twentieth century the site of the escaped bears became the location of Vogelsang's Lodges and since the 1960s has been the site of Clear Springs Apartments.

*San Marcos the River, San Marcos the Town*

In 1893 the U.S. government selected San Marcos as the site for its first fish hatchery west of the Mississippi River. The hatchery owned land on both sides of the river but placed its ponds and office on the west bank. Its old ponds now serve as landscape features on the Texas State University–San Marcos campus. A 60-foot-diameter wooden waterwheel raised water from the river into a flume that carried it across the Austin highway (now University Drive) to the ponds. The ornate wooden office building sat at the corner of what is now University Drive and Sessom Street. As part of the 1976 bicentennial celebrations, the city moved the Fish Hatchery building to its current location on the river behind the Chamber of Commerce building.

In 1899 state officials and community leaders established Southwest Texas State Normal School on Chautauqua Hill just west of the river. Just as now, students in the early twentieth

*Vogelsang's Lodges at the site of the current Clear Springs Apartments must have been a pleasant place to spend a summer vacation. Postcard is dated 1940. From the collection of Jerry and Jim Kimmel.*

century flocked to the river, although less skin showed. During 1915 and 1916 students swam in the pool below the powerhouse, but in the summer of 1916 President C. E. Evans, biology teacher and coach Dr. C. Spurgeon Smith, chemistry teacher P. T. Miller, and Dr. S. M. Sewell considered developing a swimming area below the highway bridge.

In 1916 the river split just below the bridge. The west channel was originally a millrace, but a flood moved the main flow of the river into that channel, forming an island between the old and new channels. Sewell waded around in the west channel, in water mostly 2 to 3 feet deep, the deepest being only waist deep. Sewell, later known as "Froggy," described his experiences in developing the Riverside Park to Chloe Walker Sanborn,

*The U.S. Fish Hatchery, built in 1893, was near the river, below the university's Old Main Building. The Fish Hatchery office building was moved to its current location downstream in 1976 as part of the city's bicentennial project. From the postcard collection of Jim Pape.*

*The nineteenth-century U.S. Fish Hatchery ponds are now attractive features on the university campus. The office building was moved adjacent to the river and provides a congenial meeting place for community groups.*

*This waterwheel lifted water from the San Marcos to supply the ponds of the U.S. Fish Hatchery. It was located on the right side of the river, downstream from today's Sewell Park. From the postcard collection of Jim Pape.*

*Riverside Park at Southwest Texas State Teachers College was the site of aquatic contests and performances. Photo reproduced from the San Marcos-Hays County Collection at the San Marcos Public Library.*

who wrote a master's thesis about the park at Southwest Texas State Teachers College in 1944. Sanborn relates Sewell's observations that "there was soft mud everywhere and the stream was choked with old trees and limbs embedded in the mud. The space stretching from the river to where the bath house now is was a bushy, densely weeded marsh. There were no trees save three or four big willows."

At the time, the west bank was still part of the Federal Fish Hatchery, owned by the U.S. Department of Commerce, Bureau of Fisheries, and the east bank was owned by the San Marcos Utilities Company and other private owners. Evans secured leases for 4 acres on the west bank and 17 acres on the east bank. Early in 1917 J. A. Clayton, the general yardman at the college, used a mud scraper and mule teams on each side of the channel to dig a pool in time for summer.

Clayton and his mules dredged the river, ultimately creating a variety of swimming areas, some as deep as 10 feet. Dredging necessitated bank stabilization, which first was done with a retaining wall of vertical cedar posts. By 1928 administrators realized that concrete retaining walls were necessary and received bids for twenty-two dollars per linear foot for retaining walls and sidewalks. Deciding this was too expensive, the administration authorized Clayton and his assistant Rufus Wimberley to build concrete slabs,

*Sewell Park, named for Professor "Froggy" Sewell, who founded Riverside Park in the same location, is one of the most popular places in the city and is the site of the annual New Year's Day Swim organized by the San Marcos River Foundation.*

which they then lowered into place against the old cedar pilings, at a cost of about eight dollars per linear foot.

The paved area on the east bank was once the island, the old river channel passing east of it. It was cleared in 1920 or 1921 to become a picnic ground connected to the mainland by bridges. The terra-cotta concrete slab still used in Sewell Park was poured in 1935.

Today we enjoy the pleasant walk from the old Fish Hatchery building, crossing the river on the little footbridge, to an impressive playground in City Park. This was an important but very different place for about forty years, beginning in 1907. At that time there were probably no buildings on the east side of the river. Local historian Al Lowman says the area was a bedding ground for cattle on trail drives. But the growing city faced a serious health threat posed by human waste. In 1889 the city council passed a resolution against "backyard privies that were kept in an offensive and bad condition" and "livery stables with an accumulation of decaying, rotting matter, offensive to the smell and exceedingly dangerous to the health of the inhabitants of the city." Because the land slopes toward the river, some amount of these wastes reached the river.

In 1907 the privately owned San Marcos Sewer Company built a sewerage system in San Marcos. The company also pumped water from the river to irrigate crops where Strahan Coliseum is now located. The company's sewage "treatment" consisted of a 40-foot sump and settling tank located in what is now City Park. The pretty little bridge that crosses the river near the relocated Fish Hatchery building was called "Black Pipe" because it supported the sewer pipe to the treatment plant. Effluent was released to adjacent cultivated land and allowed to decompose. Because the area was upwind and close to the city square, this was a most unpleasant solution. Also, many houses were not connected to the sewerage system. Homeowners placed their privies over drainages so they would not have to pay someone to cart away their waste, which, of course, ended up in the river.

To solve its wastewater problem, in 1916 the city of San Marcos installed an activated sludge wastewater treatment plant. The activated sludge plant was an astounding innovation, most unusual for a small town in Texas. The process had been successful in experiments but had never been put to practical use in a city, so San Marcos took an expensive risk. The activated sludge process hastens bacterial decomposition of human waste by bubbling air through the effluent. With a few modifications, the plant worked well and was a model for many cities. The sewage treatment plant was used until after the Second World War. However, perhaps by that time the plant was not very effective, because that was when the Gary

*City Park, a cattle bedding ground in the 1800s trail drives, provides residents of San Marcos easy access to their river.*

*Now much cleaner than in the past, the shallow, clear, cool water in City Park is sometimes serene, sometimes filled with activity.*

# Bathing Regatta

The fact that a small, out-of-the-way teachers college would have such a river park, offer swimming and lifesaving programs, and hold an annual water pageant was truly unique in Texas and probably the United States in the first half of the twentieth century. In the summer of 1927 the *San Marcos Daily Record* carried the following story:

*Thousand College Bathers Furnish News Reel Topic
Bathing Regatta and Red Cross Life Saving Annual Review Attract Fox News Men; Picture to Be Shown at Palace as Soon as Released*

Fox News cameramen as well as several commercial photographers and a representative of Collier's Weekly with still cameras were attracted to San Marcos Wednesday, the occasion being the bathing regatta and Red Cross Life Saving annual review in which more than 1,000 students of Southwest Texas State Teachers College took part.

Riverside, the college bathing resort, was the mecca for both students and townspeople on Wednesday—the students going betogged in bathing costumes and swim suits of every style and hue, while newspaper reporters representing out-of-town papers, still camera photographers, chamber of commerce officials, and other interested parties made up one of the largest crowds that has assembled at the college bathing resort in many months.

*1000 Take the Water*

A spectacle such as has probably never been witnessed in the State before greeted the eyes of the visitors when the more than one thousand young college students took the water in an invigorating plunge at a given signal from the director of Red Cross Life Saving, Prof. S. M. Sewell, of the college.

Riverside, the college bathing pool, has long been one of the strongest drawing cards of the college, for nowhere in Texas, nor in this country is there anything which will compare with the bathing facilities offered at the college. The pool, located in the San Marcos River, is a natural one through which many thousands of gallons of pure, crystal clear water flows every minute of the day and night. The source of the river is located within the city limits of San Marcos, and less than half a mile above the swimming pool, thereby furnishing an abundant supply of pure water and at an even temperature of about 72 degrees the year round.

It might be hard for the average person to conceive of a swimming pool which would accommodate 1000 bathers at one time, but even with this vast throng in the water, there was still ample room for as many more.

*Cameramen on Hand*

Such an event as this could not escape the ears of the Fox News cameramen, ever on the alert for that bit of news which savors of the unusual....

Some of the most notable features of the great water sports from the standpoint of the news cameraman were the grand parade of more than 1000 students, mostly girls, clad in the modern swim suits; a review of the more than sixty Red Cross Life Savers, each having passed all the strict tests necessary to win this honor; the rescue of a 280 pound man by a slight girl of 110 pounds, using the approved methods set down by the Red Cross; and many fancy diving stunts by experts.

*Rebecca Reardon Termed Miss Hercules*

Rebecca Reardon, student from San Marcos, who has for the past year been teaching at Kingsville, was the leading lady in the most spectacular exhibition of the afternoon. She won the name of "Miss Hercules" when she alone took in tow fifteen other girls and swam with them for more than twenty yards, demonstrating the practibility of one expert swimmer rescuing a number of persons who might be thrown into the water and not be able to swim ashore.

*Red Cross Life Savers Review*

More than sixty qualified Red Cross Life Savers took part in the exhibitions. These students have taken special work in this rescue work under the tutelage of Prof. S. M. Sewell of the college and have all passed all the tests incident to qualification.

Field commander prohibited his airmen from swimming in the San Marcos River, although school groups continued to do so. After the war an entirely new system was built on the Blanco River, and later another new plant was built farther downstream on the San Marcos River.

## *Old Fish Hatchery to Rio Vista Dam*

Today's Hutchison Street was originally named Mountain Street. It ended at the river where a dam and mill were built in about 1875. They were still there in 1895 but gone when Taylor conducted his survey of waterpower in 1903. As we pass under the old Black Pipe Bridge our kayaks enter a wild-looking stretch of the river where the clear water and incredible profusion of vegetation seem straight out of a jungle film. Methodist Bishop Doggett wrote about the San Marcos River to the *Christian Advocate* in 1877:

> The marvel of this wonderful river, however, is not its abrupt origin or its crystal clearness, but the wealth of sub-aquatic vegetation. Its margin is not only lined with overhanging shrubs and clustering heaps of wild tresses of long and silken grass springing from its depths and floating in the current off for twenty or thirty feet, but its entire bottom is covered with an almost unbroken tissue of delicately tinted and beautifully variegated vegetation blooming beneath the surface, under whose picturesque foliage the lithe, agile fishes perform their graceful motions, and whose crystal caves the imaginative Greek would have peopled with laughing water nymphs. I doubt if any water scene of the same extent abounds with more transcended beauty. It is a genuine, original green-house. It is nature's own conservatory, where her rarest productions are preserved in amaranthine freshness, encased in a framework of rustic grandeur, and seen through surfaces of perpetual purity. . . . One must be incurably obtuse to look into this mirror of nature and not be transported with its imagery.

The "wild tresses of long and silken grass springing from its depths" that Bishop Doggett described are Texas wildrice, which is unique to the upper San Marcos River and is listed as an endangered species under the federal Endangered Species Act. Any activity that will harm the wildrice is a federal crime, including cutting the wildrice, dredging the riverbed where it grows, or excessive pumping from the Edwards Aquifer, sufficient to stop the spring flow. Were he still alive, C. W. Wimberley would probably express a strong opinion about this management policy, considering that he observed the growth of the wildrice to be an annual "problem":

> To clear the river of moss and river grass, two farm spiked harrows were laced together on channel irons, looped at each side with chains where long cables were attached. With mule teams stationed on either side of the river, this contraption was dragged back and forth across the river bottom freeing it of all plant growth in roiled clouds of silt and wastes, doing a poor job of substituting for nature's floods which clean house by the periodic unhampered flushing of debris from its rivers.
>
> With increased stipends from the state legislature, often in response to Prexy Evans' pleas, the college bought one of Henry Ford's first farm tractors, this one turned into a dragline of sorts by a firm in Oklahoma.
>
> Equipped with a spool of cable on each rear axle, this machine could drag the river from all angles with a 12-foot length of I-beam laced with gin augers on either side. In proper hands, this operation did no harm to the river's natural black soil basin, but unfortunately could only stir the gravel, sand, and rocks man had placed there regardless of the route they traveled to get there. Dredging with "mudbucket" is an entirely different matter.
>
> Keeping the river clean in those days was a constant chore, for within a very short time the river grass and mosses were again in place waving at you from the river bed.
>
> In my unbiased opinion this dragging operation has proven to be the only practical, available means of flushing the river of its burden of

*Black Pipe Bridge, which long ago carried the city's wastewater across the river, now serves more pleasant functions.*

of the Great Plains and prairie land of the U.S. Repeatedly, I have seen the riverbed spring to life, wild rice and all, after being dragged free of the mess we have helped to create.

Our modern environmental ethic is offended by such primitive and destructive tactics, but the wildrice continues to diminish under our more sensitive care.

Removing plants from the upper river was a thriving business for fifty years. In 1924 the Federal Fish Hatchery received a letter requesting a shipment of "river moss" to use in aquaria. Lucious Parish worked for the hatchery and was given the opportunity to fill the order on his own time. He hired helpers in 1934 and resigned his hatchery job in 1944 to work full-time in the business. The Rogers family also owned an aquatic plant business, harvesting from the river. Parish's business thrived through the years, while the others ceased operation. As late as 1976 Parish employed ten helpers, each of whom harvested as much as 1,500 pounds of aquatic vegetation per day from January through May. In the early 1970s Parish built a plant at Rio Vista Dam with a 24-foot long concrete tank for separating the vegetation and refrigeration facilities. Parish's concrete tank is now under a popular riverside restaurant. Only about half of the harvest was native plant material. Parish planted a variety of exotic vegetation to meet mar-

waste, especially during the drought flow from the Edwards Aquifer.

Be that as it may, with city hall an adjunct of Old Main, it was easy enough for a learned prof to stop the dragging operation by invoking the rhetorical term "plowing the river"—branding this operation with all the sins attributed to the old moldboard plow for destroying the virgin turfs ket demand, and those exotics now comprise a substantial part of the plant community in the upper river.

The river splits a short distance downstream from Black Pipe Bridge. The stronger current goes to the right, but we can paddle our kayaks down the left channel around a heavily vegetated island to reenter the river about 600 feet downstream. The right channel was Fromme's Ditch, dug in about 1876 to irrigate Fromme's Garden, a farm lot in the original San Marcos plan. The shallow left channel is the natural channel. A dam was built at the same time to direct water down the artificial right channel. By 1895 both the island and the farm lot were owned by Walter Tips. The original town map calls the island "Hell's 1/2 Acre."

Today you can sometimes see the camp of homeless people tucked up Purgatory Creek, but during the 1930s this part of the river was a gathering place for a variety of boys and men, as C. W. Wimberley described:

> On leaving the I.& G.N. you entered the Hobo Jungle where the "Knights of the Railroad" rested among the bloodweeds, cooked their Mulligan stew in five-gallon cans, ate handouts, chewed Cooper's day-old bread, for Leslie Cooper was an easy touch to any man he deemed to be hungry. A big tree leaned low over the swift water where they bathed and washed their clothes. And during the Great Depression, many a good man was lost to this jungle life.

*Sam Jefferson hauled water for twenty-five cents per barrel from McGehee's Crossing, now Hopkins Street in San Marcos. Postcard is dated 1909. From the collection of Jerry and Jim Kimmel.*

At the other side of the circle we had Katy Hole where pore white, Negro, and Mexican kids shared the right to swim naked in the first integrated place in San Marcos and, maybe, Texas. The lower end of the pool served as a bath place for older men.

Often carrying bundles under their arms, ancient Mexican men would silently come to the lower reaches of Katy Hole where they would bathe and rinse their clothes, then sit quietly on their haunches, smoking their wisp of a cigarette while their clothes dried atop the bushes. They looked with benevolent eyes upon the young generation cavorting in and out of the water at a respectful distance.

Much of the right bank of the river downstream of the Missouri, Kansas, and Texas Railroad bridge bordered what was called Katyville, the home of most of the city's minority population. In the 1960s the federal urban renewal program acquired homes along the river and moved the residents to public housing.

The national bicentennial celebration in 1976 included grants and other incentives for cities to develop projects to celebrate the nation's first two hundred years. San Marcos' nationally recognized project was called "Beauty Along the River." The project, organized in 1973 by the Spring Lake Hills Garden Club and including civic clubs and youth groups, involved building a river walk pathway and a "playscape," restoration of the Charles S. Cock house and Rio Vista Park, initiation of the Memorial Grove, and musical events on the river. The bicentennial project also included initiation of five detention dams on tributaries of the river to reduce the effects of floods. Lady Bird Johnson made the first contribution to the matching grant. The riverfront has remained important to the city, with continued extension of the walkway, addition of lights, and interpretive signs.

Rio Vista Park is one of the most popular parts of the upper San Marcos River. Rio Vista Dam forms a long pool that keeps the water level relatively constant all the way upstream to Burleson's dam. A large city park is on the west side of the river, and a popular restaurant overlooks the river from the east bank. Almost any time of the year kayakers practice their moves in the overflow of the dam, which was strengthened and modified to include recreational rapids in 2006. But Rio Vista Dam was not built for amusement, and it was built in the midst of legal wrangling.

What we now call Rio Vista Dam was originally called the Malone-Bost Dam, after W. D. Malone and P. T. Bost, who built it in 1904. The dam's purpose was to divert water from the river for irrigation and to provide power for a mill and an electric-generating dynamo. The headworks of the diversion canal are still visible on the east side of the dam.

However, the new dam and diversion canal posed a major threat in the minds of Frank and Mattie Glover, who farmed 50 acres west of the river,

*This irrigation dam on the San Marcos River is possibly the Malone-Bost dam, now Rio Vista. The postcard is undated, but note the visitors' formal dress. From the postcard collection of Jim Pape.*

just downstream from the proposed dam, where Ramon Lucio Park is now located and a bend in the river outlines what is sometimes called Glover's Island. In 1895 the Glovers had spent $225 to dig a ditch to divert water for irrigation and to power a gristmill. They also built a small earth dam to divert water into their ditch, but it washed away as soon as the forms were removed. However, a sufficient amount of water entered their ditch to support their farming operation. In August 1904 the Glovers filed a lawsuit to stop Malone and Bost, fearing their dam would dry up the Glovers' irrigation system, but by October they had reached an agreement in which Malone and Bost would divert only half of the river's water.

Although the Malone dam served practical purposes, A. B. Rogers took advantage of the water it impounded to establish San Marcos' first river resort, Rogers Park, long before he built the hotel at the headwaters. Preston Connally, the San Marcos River boy who explored the river bottom in his primitive diving bell, the sociable Tom Sawyer to C.W. Wimberley's rough Huckleberry Finn, described the park, later renamed Rio Vista:

> Rio Vista Park was a very popular swimming hole and activity center for families and youngsters of that day. We would take a watermelon in a tow-sack and place same in the water at the edge of the bank and in the evening, after a fun-filled day of swimming, diving and horsing around, pull the melon from the water and enjoy a nectar that typified those summer days on the river.

*The people of the San Marcos River have expressed their affection for it with significant community projects along its banks.*

Rio Vista Park, formerly Rogers River Resort, continues to delight visitors.

The Rogers Park (River Resort) building is now the site of a Rio Vista Park pavilion. The concrete walkway to the island is still there. From the postcard collection of Jim Pape.

On the 4th of July, every year, a water pageant was held at the park which included bathing reviews with the latest in women's swimming attire modeled by shapely ladies on the footbridges across to the island. This was always attended with fanfare and admiration, certainly by the men.

A. B. Rogers owned the park at that time, and the facilities were outstanding, with a swing tower just below the trestle, then a high diving board next to steps down to the water's edge. Next to the steps was a huge water-elm leaning out over the water which could easily be climbed, and used as a diving platform. Next in line was the trolley tower and platform from which one could hold on to and ride down at a pretty good clip and skim one's feet on the surface until one was submerged. A counter balance would then draw the trolley back to the launching platform. The platform also served to launch divers into the water about 40 feet below. There was a diving board next to the trolley tower and one on the island across the way. The island was grass covered with willow trees on the far side, concrete retaining walls on the pool side, with a nice sidewalk along the edge. There was a top-like float with a central stem rising about three feet above the surface, capped by a wheel which one could hold onto and spin or turn the top. It drew lots of attention.

Just down river from the bathhouse was an arbor covered dance floor and jukebox which was very popular with the young people of that day.

As with most cities located on a river, San Marcos' relationship with its river has evolved. At first the river was a resource for power and water. Soon it also became a way to dispose of waste. But fairly early on, people recognized that the beauty of the river

Rogers River Resort provided water sports and lodging. Postcard is dated 1923. From the collection of Jerry and Jim Kimmel.

## Rio Vista Dam to Thompson's Islands

*The old Malone-Bost dam, now named Rio Vista. The modified dam and additional rapids provide even more excitement for river runners.*

The small John J. Stokes Jr. Park where Cape Road crosses the San Marcos River is more commonly known as Thompson's Islands. It is an "island" only because it is between the raceway from Cape's Dam and the main course of the river. By this definition, there are actually two islands, because the raceway forks, with the right fork going back into the river upstream of Cape's Road.

In her house overlooking the lower Blanco River, Kathryn Thompson Rich tells how her family's name came to be attached to a quasi-island on the river. William A. Thompson, then forty-seven years old, left his plantation north of Lake Pontchartrain, Louisiana, in 1850. Thompson was fleeing a yellow fever epidemic that killed his younger brother and other relatives.

Thompson acquired 2,600 acres on the San Marcos River and took an unusual approach to developing waterpower. Rather than build a dam to raise the water level and thus create a "head" of pressure, he had his slaves dig a raceway 1,850 feet long that carried water to a gristmill and sawmill. The raceway sloped more gently than the riverbed, so it created a fall of about 11 feet where it returned to the river. This fall was sufficient to turn an overshot waterwheel. The millstone was from the Canary Islands, and the sawmill used a circular "buzz" saw blade.

was an important resource. Code's swimming facilities at the old Burleson dam, Rogers's Rio Vista Park, and ultimately Spring Lake Park Hotel and Aquarena Springs, all built on the beauty of the San Marcos and made that beauty accessible.

Over time, the city and the university gained ownership of much of the river corridor through the city. The university's Riverside Park, although limited to student and faculty use, at least allowed access. The city's old sewage disposal site became a park.

*Kathryn Rich*, left, *a descendent of William Thompson, describes her ancestors' canal and mill site.*

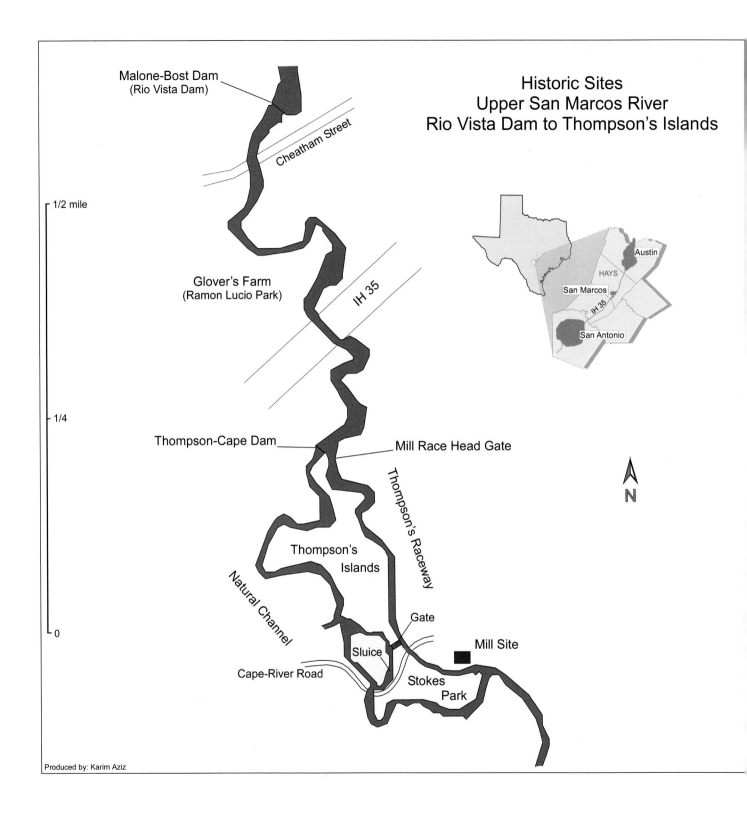

Thompson used water from the sluiceways to irrigate a large vegetable garden on the upper "island" for his family and slaves. He built irrigation flumes out of cypress logs to distribute the water over the large irrigated area.

Thompson's son, also named William, formed a partnership in 1867 to build a dam upstream to increase the velocity of flow through the raceway. This dam was built of rocks hauled from the Blanco River, held in place by a cypress framework. The rocks were so heavy they had to be hauled in oxcarts rather than wagons.

The dam came to be owned by J. M. Cape. In 1904 T. U. Taylor described it as follows: "Two miles below the head springs is the dam of J. M. Cape. This dam is constructed of framework and earth, and is 130 feet long, with a fall of 9 feet. The plant is equipped with two 48-inch Leffel turbines, which develop 78 horsepower. One-fourth mile below Cape's gin is the Thompson gin. The race is one-half mile long, and connects with the pond formed by the Cape dam. The fall at the Thompson gin is 11 feet. An auxiliary steam plant of 55 horsepower has been installed."

The area between the San Marcos and Blanco rivers was heavily forested. Pecans and walnuts were so thick that William Hardeman Thompson (the third William Thompson) told his daughter Kathryn that it was impossible to ride a horse through it even as late as 1900. Farther away from the river, out of the river "bottom," was a forest of elm trees, some with straight trunks 30 or more feet tall. Bois d'arc trees were also common.

All of these trees were valuable. The Thompsons picked pecans and sent them to New Orleans. They cut the walnut trees and sent the logs to the Texas coast for export. But they sawed the elms into planks that were the main building materials for the board-and-batten houses that soon replaced the primitive log cabins of San Marcos' first residents. The Thompson's lumber business continued until the railroad came to San Marcos in 1872, bringing the more desirable planed and milled pine lumber.

The older William Thompson operated his large plantation and built his house away from the unhealthy "river vapors," but his son lived near the mill. The Thompsons brought the first cotton gin to the area in 1850. Mrs. William Thompson named their large farm the Vail of Avoca, Gaelic for the "valley of the meeting of waters," from an English poem. Avoca is located in southeastern Ireland, although the Thompson family came from Ulster.

River work was dangerous. In 1904 William Hardeman Thompson contracted hepatitis while working in the river to build a concrete sluice gate. The city's untreated sewage entered the river less than 2 miles above Thompson's mill.

## San Marcos and Its River

San Marcos the town has used San Marcos the river for power, waste disposal, and a variety of businesses, from aquatic farming to tourism. Often the river was abused, but more and more it is venerated and protected. The city's River Corridor Ordinance is intended to limit development along the river within the city. The River Walk and city parks provide public access to much of the river within the city. Various state and federal regulations protect the river. Texas State University–San Marcos and the Texas Parks and Wildlife Department are working to convert the Aquarena Springs theme park into the education-oriented Texas Rivers Center. Ralph the Swimming Pig, sky rides, and trained chickens are gone, to be replaced by an interpretive center that tells the story of the springs and river.

Truly, you can never step in the same river twice because rivers constantly change. But the past makes the future. The early San Marcos pirate stream cut into the rocks, perhaps forming the springs. For millions of years plants and animals used the springs and river, some of them evolving into unique forms found only in this particular place. Men built dams and then changed their use. Some people planted exotic vegetation and introduced nonnative animals, while other people formed laws to protect native vegetation and animals. The river from the springs to Thompson's Islands is the most used and modified part of the San Marcos. What will our uses and plans create in the future?

# CHAPTER 6

# Dams and Towns

## A Nostalgic River Landscape

WE DELIGHT IN PADDLING our little kayaks in the upper river because the water is so clear—we Texans don't have much experience with clear water. But as we travel downstream, the San Marcos begins to look like most Texas rivers. After the confluence with the Blanco it becomes more and more murky, a pale green. This disappoints, because the clear water upstream is such a delight, but the river still has a certain charm.

The river is turbid because the water is warm and has an abundance of nutrients—all of the requirements for a profusion of life in the form of phytoplankton, which produces the pea soup tint. Some of the nutrients are the results of upstream pollution; some are naturally present. The turbid water reduces the amount of light that penetrates, so there is less large aquatic vegetation than in the clear upper river. But biologically the lower river is more productive. The phytoplankton is the primary producer, like a flowing grassland. Zooplankton eat the phytoplankton. A host of crustaceans and insects eat the phytoplankton and zooplankton. Fish eat the crustaceans and insects. And snakes, birds, raccoons, and people eat the fish. Murkiness may not be pretty, but it is very useful biologically.

The lower San Marcos runs through a landscape made up of sediments deposited by water from the west. This alluvial soil is easily eroded. Thus,

*The road in the left center is probably one route of El Camino Real. San Marcos de Neve was located on the river's right bank in 1808 near the road crossing. The rich soil of the Blackland Prairie supported a prosperous cotton industry and now produces grain crops.*

the river is able to do what rivers especially like to do—glide across the land, wiggling its hips at every curve, widening its path at the outside of the curve, and depositing material on the inside of the curve, making itself a sinuous, ever-changing path. This is natural and almost poetic, but it also means that much of the riverbank is steep and crumbly, not easily accessible nor particularly pretty. Dr. Ferdinand Roemer, the German geologist who inspected Texas in 1846, remarked on the steep banks of the middle courses of Texas rivers.

Beauty was not the point for millions of years. The river did its work. It eroded, transported, and redeposited sediment. It provided habitat. It helped carry water to the ocean.

Then, maybe twelve thousand years ago, along came humans. Did these folks think about beauty? If not, surely they thought about water to drink, food to eat, and a cool shady place on a blistering plain. Unquestionably, they used the river, but either they left few traces or the river long ago removed them or covered them with sediment as it meandered across

*At what is today's Westerfield Crossing, the Spanish residents of San Marcos de Neve may have recognized the beauty of the San Marcos River when it was calm. However, its rampaging floods contributed to the colony's failure.*

its broad valley. Among the little human evidence that survived is a specific type of Early Archaic stone point from six thousand to eight thousand years ago, called "Martindale," after the river town it was found near.

After their short-lived mission near the springs in the 1750s, the Spanish had little to do with the San Marcos River except to cross it as they traveled El Camino Real to their East Texas missions. However, in 1807 the Spanish faced a major threat from the United States. President Thomas Jefferson claimed that the Louisiana Territory, which the United States purchased from France in 1803, extended to the Rio Grande. The governor of Texas, Manuel Antonio Cordero y Bustamante, took action to establish colonies along El Camino Real, in order to protect the Spanish claim to the territory. This plan included a colony on the San Marcos River, where El Camino crossed. The governor authorized Felipe Roque de la Portilla to lead the settlement and promised him a land grant and financial support.

Portilla and his colonists arrived at the river early in 1808. The settlement was probably doomed from the beginning. Perhaps as few as ten people accompanied Portilla, although there may have been as many as fifty-two. By 1809 the census indicated seventy-three people in the colony. The records say that the governor failed to send soldiers, but local stories tell of a Spanish fort whose cedar picket remains were still visible in the late 1800s. Perhaps the colonists built their own fort to try to protect themselves and their 1,700 animals.

The river quickly made its presence known with a flood in June 1808 that destroyed most of the work the colonists had done in the first few months of the colony. They rebuilt and struggled on with personal loans from Portilla, but insufficient military protection forced them to abandon the colony in 1812. Governor Cordero neither awarded the land grant he had promised to Portilla nor reimbursed him for the money he invested in the colony.

Although the Spanish never had a strong presence on the San Marcos, by the 1830s Anglo settlers had arrived, attracted by the rich soil of the river valley. Although in some ways the new arrivals fit our romantic images of the self-sufficient pioneers, they were also participants in a global agribusiness enterprise. Stephen F. Austin, the first impresario of Anglo American settlement in Texas, gave extra land to settlers who would grow cotton and to blacksmiths and carpenters who could build cotton gins. By 1852 Texas was eighth in the nation in cotton production. Cotton was ginned and baled locally, then hauled to Texas ports, especially Galveston, to be shipped to buyers in the United States and Great Britain, making fortunes in Galveston.

Horses or mules harnessed to lever arms walked in a circle to power the gins. This horsepower required feed, but power could be generated more cheaply by small dams on the river. These little dams held back enough water to form a drop that was sufficient to power cotton gins, gristmills, and sawmills.

The rich alluvial soil that the river deposited during its meanders supported a profitable cotton industry until about 1929. Fancy old homes in San Marcos and Lockhart are evidence of the patrician life the cotton fields supported, even long after the death of the slave-based plantation culture of the South. One of San Marcos' favorite restaurants is in a huge old gin building from this time, and Luling is restoring its impressive mill and gin complex at the dam.

Boll weevils, foreign competition, federal control programs, and synthetic fibers brought about a major decline in cotton production in Texas during the early twentieth century. But by that time the towns were established on the river, their residents enjoying the electric lights powered by the water flowing over their small dams.

## *Dams, Mills, and Towns*

The river below Thompson's Islands is a remnant of a nineteenth-century landscape—almost like something out of New England, if not old England—little farming communities that grew beside their mills with their feet sometimes in the rising river, a canopy of

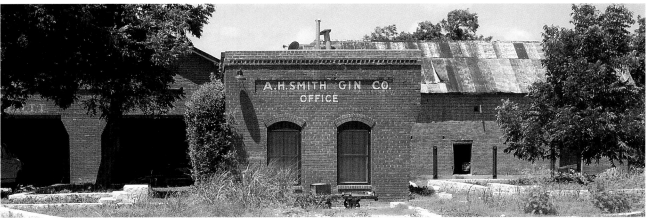

*The San Marcos River provided power to process the agricultural products of its valley. Martha Nell Holmes, center bottom, stands beside the marker of Nancy Martindale, her great-grandmother and founder of Martindale. Patsy Kimball, upper left, points out the features of the Fentress dam.*

trees overhanging a placid pool held back by an old dam beside an abandoned and dilapidated mill building with its machinery in disarray. The old mills ceased operation a half century or more ago. The populations of the small towns became smaller and turned their backs to the river. Lush riverside vegetation began to reclaim the rocks and timbers of the dams and mills, but no one needed the old mills or had the money to maintain the useless structures.

This nostalgic landscape is what you find as you seek out the riverside in Martindale, Fentress, and Ottine. Staples, Prairie Lea, and Luling are different, as we will see. But even in the towns that seem to have forgotten the river, if you look, if you talk to people, you will discover a love for the San Marcos that often goes back five generations. And you will find new generations that have joined these old families and are willing to let the river dominate their lives in ways that most modern people would not tolerate.

Floods? The modern would have the U.S. Army Corps of Engineers build a flood "control" dam. If this didn't happen, then the Federal Emergency Management Agency would buy their houses in the floodplain. Not the San Marcos river rats. Few of them want flood control dams or even build their houses on stilts. They accept floods as part of the price of being river rats. They just move the family keepsakes, dogs, cats, and kids to higher ground. The rest of the stuff is just stuff, replaceable or unnecessary anyway.

The river runners and environmentalists do not like dams and would like to see all of them removed from the San Marcos. In many ways they are right. A free-flowing river is a rare treasure today, and dams usually result in an actual loss of water to the river system. However, the dams are historic, and they gave the San Marcos River much of its human history.

## Cummings Dam

Cummings Dam, the first one downstream from Thompson's Islands, was the last one built and, because of its date and location, no town grew up alongside it. Just below the confluence with the Blanco River, the dam was started in 1905, but the project was abandoned for almost ten years. As described in the Hays County Irrigation Records Book, J. A. Bachman and Z. P. Jourdan revived the project in 1914, with the intent to back water into the mouth of the Blanco. From there they would withdraw water from the resulting lake through a canal 10 feet wide and 3 feet deep. This canal would convey water to laterals that would irrigate ten thousand acres. Ernest Cummings, who with his father, J. D., bought the dam in 1944, said that its foundation is made of bois d'arc and cedar logs placed in blue clay below the river sediment.

In addition to irrigation, the dam was built to generate electricity. Texas Power and Light Company bought the dam in 1928 and the Lower Colorado River Authority bought it in 1939.

The Cummings family bought 456 acres near the confluence of the San Marcos and Blanco rivers in 1944. The family was in the oil pipeline business and had developed innovative methods and machinery, including the bulldozer. The Cummings applied their knowledge and equipment to the challenge of modern agriculture. They pumped water from the impoundment behind the dam, using 12 inch pipes in a hub and spoke system to irrigate alfalfa fields.

The Cummings' Green Valley Farm was a prosperous showcase for irrigated farming and cattle production. However, by the mid 1970s the effluent from San Marcos' wastewater plant was so bad that it ruined their irrigation system. The family lost several hundred thousand dollars in irrigation equipment. Experiencing the polluted river personally, Ernest Cummings became one of the founding members of the San Marcos River Foundation, to defend the river from those who would harm it.

About a mile and a half below Cummings Dam is what the river runners call Old Mill Rapids. This was the dam J. C. Jones built in 1896, using a cedar timber framework filled with gravel, stones, and concrete. It had two turbines, one 50-inch diameter Leffel-Samson that produced 95 horsepower and one smaller Leffel that produced 6 horsepower. These

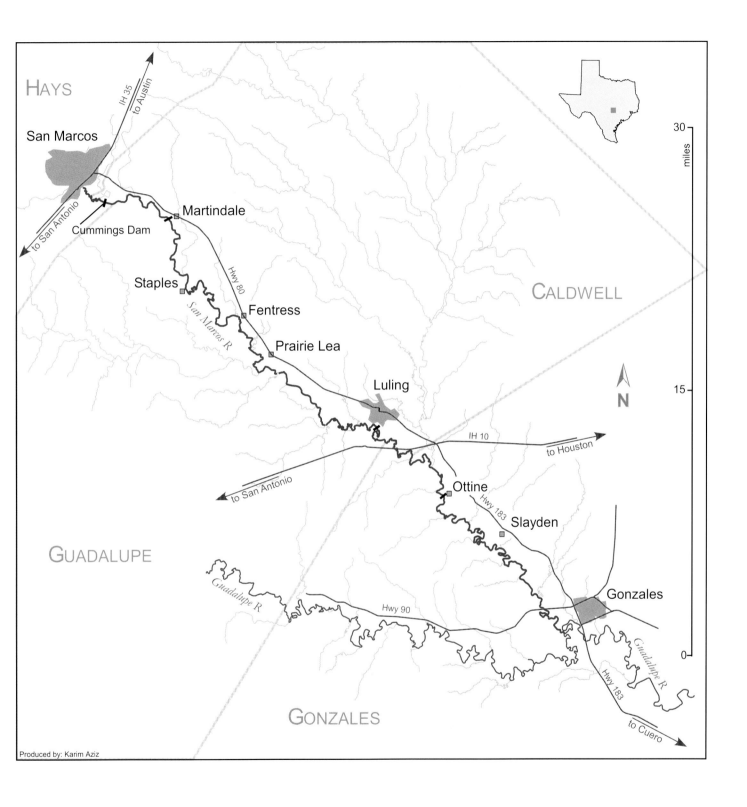

## Texas Water Safari

Imagine paddling your canoe almost nonstop for 260 miles, down the San Marcos to the Guadalupe, to the mouth of the Guadalupe, and finally across 8 miles of open water of San Antonio Bay to Seadrift. The trip is "almost" nonstop because you will have to portage around nine dams and work your way through maybe forty logjams. You leave the headwaters of the San Marcos at 9 A.M. on Saturday, usually the last weekend in June. No one on shore can give you help, food, or supplies other than water or ice. If you don't reach Seadrift within 100 hours, you are disqualified. If you want to win, you must get there in about 36 hours. The record time of 29 hours and 46 minutes was set in 1997 by Fred and Brian Mynar, John Dunn, Jerry Cochran, Steve Landick, and Solomon Carriere.

There are two types of people—those who dedicate themselves to the "toughest boat race in the world," and those who can't begin to imagine doing such a thing. The second category is much larger. But those in the first group say there is nothing else like it, and they keep coming back. Logjams, dams, water moccasins, heat, fire ants, fishing lines suspended from trees, blisters, nausea, exhaustion, and hallucinations are all part of the experience. Although it makes up only 76 miles of the total 260, the San Marcos part of the race is the most dangerous.

The Texas Water Safari started in June 1962, when Frank Brown and Bill George set out from Aquarena Springs to paddle a 14-foot Lone Star aluminum boat from San Marcos to Corpus Christi. "Aquamaids" swam down Spring Lake in front of the SMS *Aquarena* like porpoises. Frank Brown was the San Marcos Chamber of Commerce manager and conceived the trip to generate publicity for the city. Bill George owned a local restaurant.

With pith helmets, shotguns, and fishing poles, Brown and George set out in true safari style to travel the river and live off the land. It took them two weeks and six days to reach Corpus Christi. Significantly, they didn't do it again. But the next year it became a race and has continued uninterrupted since.

The race now has eight categories of watercraft, ranging from unlimited to novice nonracing canoes, but all must be human powered. There are class trophies, so you can paddle your solo kayak and perhaps win. But if you want to be the first to reach Seadrift, you must be in one of the 40-foot boats with six to eight paddlers, simply for the muscle power. Fred and Brian Myner, with other partners, have won the race thirteen times in these large boats.

---

powered a cotton gin, corn mill, and a Westinghouse DC generator that supplied electricity for thirty-five lights. The turbines also powered pumps that supplied water to irrigate 20 acres.

A little more than 2 miles below Old Mill Rapids is Cottonseed Rapids. These rapids are the remains of W. S. Smith's dam. T. U. Taylor's description of Smith's dam in *The Water Powers of Texas* indicates it was representative of other nineteenth-century dams on the river.

>About two miles northwest of the town of Martindale, W. S. Smith owns and operates a cotton gin and corn mill on the San Marcos River. His dam consists of a central portion of timber framework, filled with rocks and gravel, 7 feet high and 200 feet long, constructed by placing alternately pieces of 10 by 10-inch cypress timber longitudinally and crosswise of the dam and filling in the spaces left between these with the rock and gravel. These timbers are not laid flat upon their squared faces, but are placed upon edge, and are dapped or

*The Texas Water Safari is considered the world's toughest canoe race.*

let in several inches where they cross each other, and secured by three-fourth-inch bolts that extend completely through the dam from top to bottom. The lowest course, in which the timbers lie crossways of the dam, rests upon large longitudinal logs and extends 4 feet beyond the toe of the framework of the dam proper. Upon this extension are piled stones of all sizes up to the height of the dam, thus forming the downstream face. The top of the framework slopes backward and upstream to a slight extent, and is 16 feet in width, thus making the dam 20 feet wide at the foundation and 16 feet at the crest, the upstream side where the slope of the top ends being vertical. The ends of this wooden dam join on either side a dam of roughly shaped rock and concrete, the section next to the gin and mill being 120 feet long and that on the far side 80 feet, making a total length of dam of 400 feet. This dam was built in 1897, and affords a head of 9 feet on the turbines. Mr. Smith intends to raise the dam several feet, however, in the near future. The power is derived from three turbines—one 40-inch 1893 patent Leffel wheel, which produces 68 horsepower; one 30-inch 1899 patent Leffel wheel, giving 40 horsepower, and one 18-inch special wheel, giving 5 1/2 horsepower, making a total of 113 1/2 horsepower. This is utilized in operating the cotton gin, corn mill, sorghum mill, and a small electric-light system. The machinery consists of ten gins, two presses, one cane mill, one sorghum mill, and a 125-volt Westinghouse direct-current dynamo. The total cost of the plant was $25,000.

At the beginning of cotton production in the San Marcos valley, it was common for gin operators to shovel the cottonseed into the river—probably giving Cottonseed Rapids its name. However, due to the high quality of cotton produced in the valley, the seeds later were kept and sold throughout the cotton-producing regions of the country.

## Martindale

Martindale is such a classic example of a southern farming trade town that it provided the setting for movies like *A Perfect World* and the *Newton Boys*, not to mention scenes from *The Texas Chainsaw Massacre*. And Martindale is truly a river town. Main Street is just a little more than 400 feet from the river, and only about 17 feet above the river. Thus, the river frequently visits Martindale.

Martindale is named for Nancy Martindale, who donated land for the town in the mid-1850s. She and her husband moved to Texas from Mississippi in 1851, but her young husband soon died. She not only raised their nine children, but grew prosperous and gave land for a town that would uphold her moral convictions. Her gift of land stipulated that the property of any resident who gambled or sold liquor would revert to the Martindale family. Folks in Martindale still take this seriously.

Martindale was a trade center for the rich agricultural lands in the river valley. By the 1890s Martindale had four gristmills and gins and four general stores. J. W. Teller owned the dam at Martindale, 250 feet of timber and rock, similar to Smith's dam upstream. Built in 1893 at a cost of twenty thousand dollars, this dam and mill complex was one of the largest on the river, with 61-inch and 35-inch turbines, when Taylor did his survey in the early 1900s. It powered five gins and one cotton press, a corn mill, and a sorghum seed thrasher. It also powered a water pump to supply water for the town and a Westinghouse 125-volt DC dynamo for electric lights.

The Martindale cotton gin was the largest in the world used exclusively for seed cotton. There were five individual "stands," each producing seed cotton for a separate company. A. H. Smith built an impressive brick gin building in about 1910 and probably at the same time replaced the wood and rock dam with the concrete dam that still forms a beautiful mill pond at Martindale and provides river runners so much excitement.

The gin closed in 1944. As an agricultural community, Martindale peaked in 1957 with a population of 600, before cotton production shifted to irrigated land in west Texas, and the San Marcos valley shifted to grain crops and grazing. Martindale's population declined to 210 by 1982. However, as the Austin-San Antonio region grew rapidly, Martindale found new life as a bedroom community and the Texas State Demographer estimated its 2004 population to be about 1,000 people. For law enforcement, the unincorporated town depended on Caldwell County sheriff's deputies posted in Lockhart. Among students at the nearby state university, Martindale had a reputation as a congenial place to live due to the relative absence of law enforcement. This prompted the residents to incorporate as a city in 1982.

Martha Nell Holmes, Nancy Martindale's great-granddaughter, was the first mayor of the newly incorporated city. Flooding and wastewater disposal were major problems, especially in the southeast part of the town. The city of Martindale received a federal grant in 1987 to provide wastewater collection, and the Federal Emergency Management Agency purchased frequently flooded houses after the 1998 flood.

Although located on the river, Martindale does not take its water from the river. The privately owned Martindale Water Supply Corporation draws water from deep wells in an aquifer that recharges from the north.

## Staples

Possibly due to its geography, Staples even today is a little different from some of the other river towns. It sits on a rise about 50 feet above the river, which protects it from most floods. Its low dam is intact, and the riverbanks are maintained as parts of private residences.

Settlers first moved to the Staples area in 1852. Perhaps as early as 1855 Leonidas Hardeman built a dam on the San Marcos River to provide power for a cotton gin. Power from the dam was also used to grind corn,

*From the air, Martindale's intimate relationship with the river is evident.*

*A broad floodplain separates Staples from the San Marcos River.*

saw lumber, and grind sugarcane. Al Lowman's ancestors, who developed waterpower at Staples, told him that cane mills were so nasty and attracted so many flies that they lost their taste for molasses.

Confederate Governor Edward Clark established Camp Clark near the river in Staples as a training camp in 1861. The 4th Texas Infantry and Colonel Peter C. Wood's 32nd Texas Cavalry trained at Camp Clark.

Staples did not begin to grow as a community until after the Civil War. Confederate Colonel John Douglas Staples established a store about 500 yards from Hardeman's dam in 1871, and the town was called Staples Store until 1891, when the Post Office Department shortened the name.

When Taylor conducted his waterpower survey, he found that Q. J. Lowman had rebuilt the dam in 1899 and was rebuilding the mill complex, which had burned in 1901. Brothers Quincy and Roston Lowman established the Staples Water Power Company in 1890. Taylor described their renovations and improvements:

> The timbers of the old mill were unhurt by the fire, and are to be used with the new machinery. They consist of one 66-inch Morgan Smith wheel, which develops 82 horsepower; one 42-inch McCormick wheel, giving 82 horsepower, and one Leffel 26-inch wheel, giving 12 horsepower. The power is used at present in operating the waterworks of the town of Staples, but will be utilized for ginning purposes, the operation of a corn mill, and electric lights for the town upon the completion of the mill. The machinery will consist of eight cotton gins, one 1,000-volt alternating-current dynamo of 750-light capacity, a Gould pump with a capacity of 75 gallons per minute, and a 50–horsepower Atlas engine and boiler; also a corn mill. The total cost of the plant will be $25,000.

Staples was indeed moving into the modern world, with the large new-fangled alternating current dynamo that could provide electric lights for the entire community.

Staples thrived for many years as an agricultural trade center. Although the new state highway bypassed the town in 1929, the gin continued to operate until 1981. Today Staples is a well-kept little community, home to agricultural families and people who work in the Austin-San Antonio metropolitan region.

## Fentress

Fentress today is a graphic testament to the aggressiveness of river bottom vegetation. Trees, vines, grasses, algae: deep green shrouds the town. Like Martindale, Fentress lives close to its river—only about 500 feet from the river and 14 feet above it.

The recorded history of Fentress is something of a chicken-and-egg mystery. Some sources say the Cumberland Presbyterians established a church near the San Marcos River in 1869 and a settlement called Riverside developed around it. In 1870 Cullen Smith and Joseph Smith built a horse-powered cotton gin. In 1879 they built a dam and converted the gin to waterpower. But other sources say a water-powered gin was built first and the church came later. Since a gin attracts people, and a church needs people, that story is more plausible. Still, the Calvinist Presby-

*The town of Staples was first called Staples Store. This photograph was made in 1900. Photo reproduced from the San Marcos-Hays County Collection at the San Marcos Public Library.*

terians may have been predestined to be there first! But the local history teacher during the 1940s taught her students that Riverside was about 2 miles southwest of where Fentress is now located. So Fentress' past is quite uncertain.

Regardless of the sequence of conception, the good folks of Riverside applied for a U.S. Post Office in 1882 but were denied because there was already a Riverside, Texas. The town was renamed Fentress after James Fentress, the town's doctor, who also owned a large amount of land. By 1896 Fentress had about 150 residents.

When Taylor, the engineer, visited Fentress in the early 1900s, he described the dam as a timber framework filled with rock, like most of the other dams on the river. However, its wooden framework supported 2 x 12 "flashboards" that could raise the height of the dam from 6 feet to 8 feet in order to increase the power generated by the Morgan Smith turbine. At 6 feet the turbine generated 35 horsepower, and at 8 feet it generated 55 horsepower.

The turbine powered seven Munger gins, two pumps, a 115-volt DC dynamo, and a corn mill. However, since the waterpower was not sufficient to run all of this machinery at full capacity, a 50-horsepower Atlas steam engine provided additional power.

But the dam at Fentress today is not the one Taylor described. Instead of wood and stone, the massive structure is made of poured concrete. The

*Dams on the San Marcos provided essential power for the region's economy between the mid-nineteenth and mid-twentieth centuries. Examples include Cummings, top, Staples, center, and Zedler's at Luling, bottom.*

adjacent gin building is also made of mostly poured reinforced concrete, like a fortress, rather than the usual wood or corrugated sheet metal of many old gins. Clearly, this was a thriving operation. As we scrambled over the dam with fifth-generation Fentress resident and proud river rat Patsy Kimball, we found inscribed in the concrete the date "7–25–25," the height of the cotton industry in the San Marcos Valley.

Fentress of the past was not only an agricultural trade center like Staples but also a popular resort center. As in San Marcos, Fentress residents recognized the recreational value of the San Marcos River. In 1915 Josh Merritt and C. E. Tolhurst built a swimming and camping resort just downstream from the dam, complete with a waterslide and screened tents. After two years of operation, Merritt and Tolhurst sold the resort to J. C. Dauchy, W. R. Smith, and J. M. Dauchy. These partners built a skating rink that was used as a dance floor on alternate nights, to the consternation of church members.

The resort was a major attraction until World War II, when there was insufficient manpower to keep the lake behind the dam clear of debris. Ultimately the lake filled in, and the river bypassed it in the late 1940s, forming an island of about 3 acres. The resort closed in the 1970s, and the skating rink building was moved to Lockhart where it was to house a museum, a project that never materialized.

*The Fentress River Resort offered swimming and roller skating. The postcard is not dated, but this style of men's swimming suits was common through the late 1930s. From the postcard collection of Jim Pape.*

Oil was discovered near Fentress in the 1920s, and the town's population grew to 500 people by 1929. At its height in the 1920s Fentress had four general mercantile stores, two drugstores, two garages, two gas stations, a bank, a newspaper, a barber shop, a blacksmith shop, a confectionery, a hotel and boardinghouse, and a gin.

Highway 80 was constructed in the late 1920s and bypassed the old route through town. By the 1940s the population had declined to 250, and the local school was closed as the school system consolidated with Prairie Lea's. Today Zapata's Upholstery and Grocery store is the only business in town. The last official population count was 85 people in 1990, but a campground and RV resort built in 1990 continues to attract people to the river.

## Prairie Lea

Its name is appropriate. Prairie Lea seems more like a prairie town than a river town. It stretches along Highway 80, which apparently follows the path of the original route through town. Unlike Fentress, a town with its feet in the water, Prairie Lea is about $4/10$ of a mile from the river and 46 feet above it.

This little town is the oldest settlement in Caldwell County and the oldest on the river below San Marcos. The land was granted to Joe Martin of Gonzales by the Spanish in 1820, but Edmund Bellinger, the first settler, did not arrive until 1839. Bellinger was a veteran of the battles of San Jacinto and Plum Creek.

What a beautiful name, and what a mystery. There are at least three stories about its origin. The most romantic story is that Sam Houston named it after Margaret Lea, the beautiful twenty-year-old he married in 1840 when he was forty-six and whom he remained married to for the rest of his life. Although Sam Houston was a famous carouser, Margaret converted him to be both a Baptist and a teetotaler, so she certainly deserves to have a town named for her.

A less romantic explanation of the town's name is that a Dr. Lea lived there until he left to fight in the Civil War. Although he survived the war, he apparently never returned to Prairie Lea.

Some say the town was named for its geographic characteristics. It is on a slight rise on the prairie, and "lea" in Old English means open ground, or perhaps a hill.

Slave owners settled the countryside, and Thomas Mooney built a dam and mill complex on the river. A log cabin used for both the school and church was built in 1848, and Sanford Callihan opened a store in 1849.

The slave-based cotton industry provided a strong economic base for Prairie Lea. In 1852 there were two stores, a post office, a hotel, a school, and the Masonic Order's Prairie Lea Academy. In 1860 the Masons formed the Prairie Lea Female Institute.

Prairie Lea men fought in several Civil War campaigns, but none more disastrous than Confederate General H. H. Sibley's effort early in 1862 to capture the upper Rio Grande country. After Sibley's army was defeated by federal troops from California, men from Prairie Lea were stranded and starving in the desert. In June 1862, the community organized a rescue operation that saved some of the men. But Prairie Lea continued to be such a violent place during Reconstruction that some families moved to Mexico.

Life had settled down by 1877, but that year the town's seventeen stores burned. Nevertheless, the community continued to grow. Between 1884 and 1914 the population increased from 100 to 350 people.

Thomas Mooney's early dam had been replaced by J. J. Jones' in 1896. T. U. Taylor described the dam as the standard timber framework filled with rock. The 61-inch Alcot turbine produced 45 horsepower and powered a gin, corn mill, and a 110-volt electric dynamo. Nothing remains of the dam today.

Prairie Lea continued to be a center for education. The Prairie Lea Independent School District was formed in the 1940s, incorporating the schools from both Fentress and Stairtown.

Less than 2 miles southeast from Prairie Lea on Highway 80 is Stairtown, named for Oscar Fritz Stair, originally from Alabama. Perhaps because of his Deep South accent, the name of the town is still pronounced "startown."

Stairtown has no relation to the river. Although it began as a farming community, it is less than a mile from the Rafael Rios No. 1 oil well that wildcatter Edgar Davis brought in on August 9, 1922. What came to be called the Luling oil field made Stairtown into an instant boomtown. However, once the Luling oil field was fully developed, there was no longer demand for large numbers of workers, and Stairtown began to decline. Only a few houses remain today.

## Luling

Luling is known for its Watermelon Thump festival, the smell of its high-sulfur oil field, and the stories about its name. But in the future it may be most famous for its restored mill on the San Marcos River.

By looking at how its streets are oriented with the tracks, you can see that Luling is a railroad town, not a river town. Anglo American settlers moved into the area in the 1840s. Their farms were north of present-day Luling on Plum Creek and in a small community called Atlanta about 5 miles east of Luling. But in 1874 the Galveston, Harrisburg, and San Antonio Railroad built a line from Columbus to a point 3 miles west of Plum Creek. This end of the line was a classic break in transportation and became the community of Luling. People moved into Luling from Atlanta and the Plum Creek farms. With its railroad connection, Luling became the northern end of a freight road to Chihuahua, Mexico. The Galveston, Harrisburg,

*Zedler's Mill in Luling has become a popular location for fish fries, canoe races, and other festivals.*

*Zedler's dam and mill in Luling comprised a state-of-the-art agricultural processing complex. Postcard is dated 1910. From the postcard collection of Jim Pape.*

Crossing the San Marcos River just south of Luling on Highway 80, you can see Fritz Zedler's dam and mill. The first dam was built at this site in about 1874 by Leonidas Hardeman and James and John Merriweather. Fritz Zedler, Bob Innis, John Orchard, and J. K. Walker purchased the dam and associated equipment in 1884. In 1904 Taylor described the dam as the usual timber cribbing filled with stone and gravel. This dam generated 90 horsepower through a 66-inch Leffel turbine and 60 horsepower through a 54-inch Alcot turbine. These turbines powered a cotton gin and gristmill and apparently made a very nice living for Zedler, judging by his large house nearby. The reservoir formed by this dam also provided the water supply for Luling. Zedler replaced his original dam in 1914 with the present concrete structure.

and San Antonio Railroad came under the control of the Southern Pacific Railroad in 1881 and operated as part of its Sunset Branch.

The most colorful story of the origin of the town's name is that it comes from a Chinese launderer named Lu Ling, who was attracted to the town by the railroad. But the most likely origin of the name was that the maiden name of the railroad financier's wife was Luling.

Even though Luling is more a railroad town than a river town, the San Marcos River has been important to the city. In 1904 Taylor described a timber and rock dam 3 1/2 miles upstream of Luling. In addition to powering the usual sawmill, gin, and gristmill, this dam provided electric lights for the city. The dam was built in the 1870s and rebuilt several times before Taylor inspected it. He observed that its foundation was the muddy riverbed, which probably explains why, as we drift our kayaks down that part of the river, we find no trace of the dam.

*Located on the river to harness its power, Zedler's Mill also was vulnerable to the river's floods. From the collection of Randy Engelke.*

Edgar Davis' oil strike near Stairtown in 1922 turned Luling into a boomtown and Davis into a millionaire. Davis was a generous man. He gave Luling a site for a hospital; established the Luling Foundation, an agricultural experiment station still in operation; and built parks. The park on the river featured a bathhouse of reinforced concrete to withstand floods. To celebrate its opening, Davis gave the citizens a barbecue dinner, complete with bottled soft drinks. Many patrons threw their bottles into the river, thus spoiling the swimming hole with broken glass. The bathhouse went unused and deteriorated to the point that the city decided to tear it down. However, its reinforced concrete structure not only resisted the river, but the demolition crew as well. The city simply buried the old bathhouse, but the 1998 flood uncovered it. The city restored the bathhouse, which, alas, now serves the swimming pool rather than the river.

The city of Luling has recently acquired the old Zedler mill and homesite and is currently restoring it as a museum and recreation facility. The site comprises more than 9 acres and contains an amazing complex of structures and machinery that were state of the art for waterpowered agricultural processing in the late nineteenth and early twentieth centuries. The site has great potential to tell the stories of the struggles and inventiveness required to make a living from the land and water. Although Luling was not initially a river town, the San Marcos River is becoming an important focal point for this city of more than five thousand people, as it works to provide a sense of history and pride for its residents and attractions for visitors.

## Ottine

Driving south on U.S. 183 2 1/2 miles south of Interstate 10, turn right onto Park Road 11, which follows the path of the nineteenth-century San Antonio and Aransas Pass Railroad. In less than ½ mile you will see a pullout on the right. The red rocks give it the name "Red Hill," and it overlooks the wide San Marcos valley. The river is ¾ mile to the west and 132 feet below.

But if you believe in Bigfoot-type beings, probably you should not stop there at night. Local legend has it that the Ottine Swamp Monster, "The Thing," tries to push lovers' cars over the cliff.

Ottine Swamp, the monster's home, is less than 2 miles south, on the San Marcos River. The swamp, a woody form of a wetland, consists of overflows from the meandering river, a high water table, and small sulfur springs adjacent to the river. Before the pumping of water and oil in the 1950s lowered the water table, there were warm springs and mud boils. The Gonzales Warm Springs Foundation for Crippled Children established a hospital at Ottine in 1937 to treat polio victims with the 98.5°F water from the springs.

Most of Ottine Swamp is now in Palmetto State Park, which was developed in the mid-1930s. The Civilian Conservation Corps built park facilities from the native red sandstone. The magnificent old pavilion was recently designated by the National Park Service as one of the most outstanding park buildings in the nation. Today you can see where the flood of 1998 lifted the wooden roof structure and deposited a 4-inch log between the roof and the masonry wall.

Palmetto State Park is a threatened biological jewel. Hundreds of plant species and more than 240 bird species have been identified there. Here Scottish botanist Thomas Drummond discovered the phlox plant that came to be named for him and has been used as an ornamental in the southeastern United States and Europe for decades. This biological wonderland is threatened because the water table has dropped and the wetland is drying up. An old hydraulic ram pumps water into part of the wetland but not enough to ensure its survival.

Ottine is the last community on the San Marcos before the river meanders to its junction with the Guadalupe River near Gonzales. Ottine was named in 1879 for its founder, Adolf Otto, and his wife, Christine. Otto built a gin powered by springs

Randy Engelke and Jim Kimmel demonstrate the size of one of the Leffel waterpower turbines at Zedler's Mill (top center). The Zedlers operated state-of-the-art gins and mills in Luling and Ottine.

*The heavily wooded area at Ottine hosts Palmetto State Park—and maybe the Ottine Swamp Monster.*

in the swamp. Ottine was a stop on the San Antonio and Aransas Pass Railroad, and by 1897 the town had two general mercantile stores, a lumberyard, sawmill and gristmill, and a blacksmith shop. Representing both sides of the Texas personality, it also had a church and a saloon.

Adolf Otto built the first dam, but Fritz Zedler's son Berthold, from Luling, bought the dam and mills from him in 1894. Taylor reported in 1904 that the dam was the usual wood cribbing and stone dam, and powered a 72-inch Alcot turbine, producing 70 horsepower that ran a cotton gin and gristmill.

Berthold Zedler apparently did well with the dam and mills. He built a fine three-story house in 1907 in which his granddaughter Antoinette May still lives. In 1911 Zedler built a new concrete dam with two large turbines. The dam cost $8,076, which today would be about $150,000. The turbines turned massive pulleys that transferred power to the gin and sawmill about 80 feet above the river. Before World War I as many as five railroad car loads per day of walnut logs were shipped from Ottine, much of it bound for Germany. The local story is that the Germans used the walnut for gunstocks.

Ottine's population reached two hundred in 1915 but fell to one hundred ten years later. After the Warm Springs Foundation for Crippled Children was established, Ottine's population grew again to two hundred by 1946. However, the population declined again to one hundred by 1965. The Warm Springs Foundation recently closed, and only a few families live in Ottine today.

But the historical population figures do not include the Ottine Swamp Monster. Ed Syers in "The Thing in Ottine Swamp," in *Ghost Stories of Texas* wrote,

> Jackson named several others who had encountered the thing, from Luling and Gonzales, Texas. Two men, Billy Webb and Ab Ussery, running a trotline one night, saw an expanse of bloodweeds along the riverbank moving as if something large were passing through it, following them. "A big light, and they weren't twenty feet from the bank," but still they saw nothing.
>
> The "Thing" seems to be attracted to vehicles parked on Lookout Hill, which lies near the entrance of Palmetto State Park. Two young men, Brewster Short and Wayne Hodges, readying themselves to go home from a hunting trip, claimed that something unseen reared up on the back of their

*Berthold Zedler replaced the old rock and log dam at Ottine with a new concrete dam in 1910. From the collection of Antoinette May.*

*The "new" dam at Ottine powered Berthold Zedler's gin and sawmill and was a point of pride for the family and the community. From the collection of Antoinette May.*

car. They fled hastily, leaving their dogs (which had started howling) behind. This experience so disturbed Wayne that he moved into his parents' bedroom.

Lamar Ryan, Berthold Jackson's cousin, was parked one night on the hill with his fiancée, when something started shoving his pickup towards the edge of a steep drop-off. Lamar jumped out, but he could see nothing, despite the moonlight. He and his wife-to-be left hastily.

Jackson's son spoke to a couple who lived in a trailer house near the swamp. He was informed that something had once or twice shook the trailer "like a box," and that the couple had come home one day to find the wife's best dress, which had been on a clothesline, "torn in half, and each half rolled in a ball and stuck under each bed."

Berthold Jackson concluded: "I spent fifty-four months in World War II—Saipan to Okinawa—and the only time in my life that my hair has stood straight on end is when I've looked right at it and couldn't see a thing."

Berthold Jackson's cousin, Antoinette May, told us that Jackson said he once saw "The Thing" with a young one.

## Slayden

Slayden is the last community in the San Marcos valley, but it is a mile by road from the river and was not founded as a river community. It was

*Berthold Zedler's cotton gin at Ottine was one of many that operated throughout the San Marcos River valley until the 1950s. From the collection of Antoinette May.*

established in the late 1880s as a station for the San Antonio and Aransas Pass Railroad. Named for Congressman James Luther Slayden, by 1914 the town had about two hundred residents and even had telephone service. However, during the Depression and the Second World War people moved away from Slayden, and the population declined to fifty in 1946 and fifteen in 1990.

## Gipson's Ferry

James Gipson operated a ferry crossing on the San Marcos River just west of Gonzales beginning about 1839. In 1846 Dr. Ferdinand Roemer wrote, "We were ferried across the San Marcos a few miles beyond Gonzales, which is here a narrow, sluggish, muddy stream, scarcely twenty paces wide. Later we learned to know it again in its upper course as a beautiful, rapidly flowing stream of incomparable clearness."

## Covered Bridge

We associate covered bridges with New England or Madison County, Iowa, and assume that their purpose was to prevent snow from piling up on the bridge. However, John Mooney opened one on the lower San Marcos in 1856, about 2 miles west of Gonzales and 2 miles upstream from the confluence of the San Marcos with the Guadalupe River. Snow surely was not a problem, but the bridge be-

came a visitor attraction, an exciting one at that. T. Lindsay Baker, in his book *Building the Lone Star*, quotes the July 14, 1869, issue of the *San Antonio Express*:

> Anyone who has traveled by way of Gonzales, will have a lively remembrance of the San Marcos Bridge—it was one of the sensations we treated strangers to upon their introduction to Western Texas. Imagine a long narrow frame, boarded in on the sides, roofed over head, and a floor under foot, perched upon rotten poles forty feet high over a fearful, swift, deep stream. The polite stage drivers generally requested that the gentlemen walk over, while he undertook the perilous experiment of driving over the stage with the ladies inside, if there happened to be any on board.

*In 1856 John Mooney built a covered bridge on the San Marcos River west of Gonzales, about two miles upstream from the river's confluence with the Guadalupe River. From the Gonzales County Archives.*

The bridge structure was really stronger than "rotten poles." Most of the weight of the bridge was borne by two massive cut stone piers, which are still standing. However, the span between the piers and the approaches to the bridge were in fact rather spindly wood beams and piers. The covered bridge continued in use until 1902, when Gonzales County opened a new steel truss bridge.

*The old iron bridge near Slayden is still in use.*

CHAPTER 7

# Past Creates Future

*R*IVAL comes from the Latin word that means someone living across the river. Rivers, especially the San Marcos, continue to be sources of rivalry and conflict. There is rivalry over the Edwards Aquifer, the main source of the river's water. There is rivalry over the use of the water once it is in the river and over the use of land that drains into it. There is rivalry over the plants and animals in the river. No wonder Mark Twain said that whiskey is for drinking and water is for fighting.

Because a river is an integrated land and water system, its various issues are all interconnected and consequently more complex. The very flow of the San Marcos is affected by irrigation in Uvalde County and evaporation from swimming pools in San Antonio. The quality of water in the river is affected by development in the Hill Country and farming in the Blackland Prairie. Because the river contains four or maybe five endangered species, its management is a federal issue.

All of the specific issues ultimately come down to water quantity and quality. The river must have water to be a river, and that water must be of sufficient quality to support the needs of nature and humans.

*Spring Flow*

Two of the foremost scientific experts on the San Marcos, Dr. Glenn

*Without flowing water, there would be no river. Although the San Marcos River now flows steadily, it is subject to many conflicting priorities.*

Longley of the Edwards Aquifer Research and Data Center and Dr. Randy Moss, formerly of the Texas Parks and Wildlife Department, have said the greatest threat to the San Marcos River is diminished flow from the springs. Without the springs there will be no river.

Most of the San Marcos' average total flow comes from the Edwards Aquifer via the headwater springs. The water in the aquifer enters as surface runoff from the Edwards Plateau and as subsurface spring flow from the Edwards-Trinity aquifer. These aquifers discharge and recharge quickly, because their flow is directly related to rainfall. Thus, in drought time the flow diminishes. Since the San Marcos Springs are at the lowest point of the aquifer, they have never ceased to flow in historical time. In contrast, during the 1950s drought Comal Springs, at the second highest elevation on the aquifer, did cease to flow. Because much more water is pumped from the aquifer now than fifty years ago, it is entirely possible that the San Marcos Springs could dry up and the river would virtually disappear except when storms produce runoff.

The river's endangered species are under the jurisdiction of the federal Endangered Species Act, which is administered by the U.S. Fish and Wildlife Service. The U.S. Supreme Court required the U.S. Fish and Wildlife Service and the state of Texas to create the Edwards Aquifer Authority (EAA) to enforce the Endangered Species Act related to waters from the Edwards Aquifer. The EAA has the responsibility to regulate pumping to ensure that, among other major goals, flow from the San Marcos Springs is adequate to maintain the habitat for the river's endangered species.

The Endangered Species Act provides for what is called an "Incidental Take Permit," which offers a way to balance the often conflicting goals of resource use and species protection by allowing minimal and unavoidable impacts on the species. Application for the permit through the U.S. Fish and Wildlife Service requires an acceptable Habitat Conservation Plan (HCP). In March 2005, the Edwards

*The San Marcos River is influenced by its large tributary, the Blanco, and how the land is managed on the Edwards Plateau, which contributes both surface water runoff and groundwater recharge.*

Aquifer Authority issued its "Draft Edwards Aquifer Habitat Conservation Plan." The mission statement calls for development of a long-term regional HCP that will optimize use of the Edwards Aquifer while:

1. Minimizing and mitigating negative impacts upon federally-listed species dependent upon springflow from Comal and San Marcos Springs through aquifer demand management, springflow protection, and other management strategies; and

2. Diminishing the negative impact of the plan on the regional economy and economic interests of all the stakeholders.

Points 1 and 2 of the mission statement sound good, but they are on the opposite sides of a line drawn in the sand between conservation and economic development. The 172-page "Draft Edwards Aquifer Habitat Conservation Plan" provides substantial detail about ways to accomplish the conservation goal. (See appendix 3 for summary.) However, it provides virtually no guidance on how to accomplish the second goal, other than to state that "the USFWS agrees that, prior to undertaking or attempting to impose any action or conservation measure, it shall consider all practical alternatives to the proposed conservation measures and adopt only that action or conservation measure which would have the least effect upon the economy and lifestyle of the Authority and permittees, while at the same time addressing the unforeseen circumstance and the survival and recovery of the affected species and its habitat."

Of course, the Edwards Aquifer Authority and the U.S. Fish and Wildlife Service are not economic management agencies, so we should not expect them to provide detailed recommendations for economic development, as they did for species conservation. However, I would argue that "diminishing the negative impact of the plan on the regional economy and economic interests of all the stakeholders" is at least as difficult as protecting the habitat of the endangered and threatened species. Failure to develop strategies for the second goal will ultimately retard efforts to accomplish the first.

The issue of habitat conservation extends far beyond the springs or the San Marcos River. Texas journalist Michael Berryhill recently wrote, "It would not be an exaggeration to say that the Guadalupe River has kept the whooping crane from going extinct, and that the crane's future depends on whether the river continues to deliver fresh water to San Antonio Bay." San Marcos Springs and the San Marcos River provide an important part of the flow of the Guadalupe River, so the Edwards Aquifer Authority has to manage the aquifer to protect endangered species at both ends of this complex hydrologic system, not to mention the many other species of the coastal estuaries.

But the issue is even more complex and subtle than a Supreme Court ruling or sophisticated management practices. Recharge of the aquifer is affected by land use and management on the Edwards Plateau. Residential development increases runoff by decreasing the ability of the landscape to absorb rainwater. Increased runoff results in rapid, high volume flows down streams, so rapid that relatively little of the water is able to flow into the streambed cracks that are the main recharge route for the aquifer.

Brush management on the Edwards Plateau is another subtle but very important factor that affects flow in the San Marcos River. There were many fires when the Edwards Plateau was Indian land. The tall grass provided fuel. Lightning set fires, as did the Indians for hunting. Fire restricts juniper trees to ravines where they are protected and to mountaintops where there is insufficient soil to grow much grass for fuel. Live oak trees are not as vulnerable to fire as junipers and grow in clumps called *mottes*. Thus, during Indian times the Edwards Plateau was a savanna of grass, scattered live oak mottes, and juniper in the roughest places.

When the ranching cultures arrived, they overgrazed the grass fuel and put out fires when they could, so the juniper flourished. Ranchers worked to control the juniper, because it displaced grass. They cut it

by hand, burned it, poisoned it, and "chained" it by raking it over with anchor chains stretched between two bulldozers. But by the mid-1970s ranching was such an unprofitable activity that ranchers could no longer afford to control juniper. Scientists predict that unless major efforts are made to control juniper, the Edwards Plateau will become a "closed canopy" juniper forest.

So what? To use an old cowboy term, juniper "plays whaley" with the natural hydrology. First, junipers intercept as much as 80 percent of the rainfall in their thick foliage, thus preventing the water from reaching the ground. Second, all trees, including junipers, draw water from the soil and release it into the air in the process of transpiration. A 10-foot juniper will release about 3 gallons of water per day. This water is removed from the system and does not go into the aquifer, does not come out of the springs, and does not flow down the river.

*Naturally clear water is the hallmark of the upper river, but it could be permanently clouded by inappropriate human management of the Edwards Aquifer and the watersheds that drain into the San Marcos River.*

A sneakier activity of junipers is probably more harmful. The leaflets from junipers are acidic and retard growth of vegetation around the trees. This exposes the thin soil of the region to erosion and also prevents formation of new soil. Ultimately, as is demonstrated at the Texas Parks and Wildlife Department's Kerr Wildlife Management Area near Hunt, Texas, there is nothing left but bare caliche—soil particles cemented with calcium carbonate. This caliche cannot absorb moisture and causes the water to run off quickly, resulting in the same effect that the roofs and concrete of development create. J. David Bamberger removed the invasive junipers from his Bamberger Ranch Preserve and the springs flowed again, but it cost him a fortune.

The Edwards Plateau, the source of the San Marcos' water, is virtually all privately owned rangeland and is likely to remain so. Unless a landowner is rich and willing to spend his or her money, blood, and sweat removing juniper, there is no rational economic incentive to do so, and juniper will continue to impact the hydrology of this complex system. In some cases removing juniper actually violates the Endangered Species Act, which not only attempts to protect the endangered species in the river but also certain endangered songbirds that nest in the juniper. So the issue is indeed a complex and contradictory one.

## San Marcos River Foundation

**Denny Thomas, publisher of the *San Marcos Record* newspaper during the mid-1980s, wrote about the need for an organization to help protect the river. The first board of directors, including river activists and community leaders, convened in 1985.**

**The San Marcos River Foundation (SMRF) influenced creation of the city's River Corridor Ordinance and Edwards Aquifer Protection Regulations. Its members took firm positions with the city regarding the municipal wastewater treatment plant and the bed and banks controversy. SMRF also successfully challenged the Texas Parks and Wildlife Department regarding effluent from the department's San Marcos Fish Hatchery.**

**SMRF has generally avoided the strident tone of some environmental organizations and has consistently relied on credible scientific studies to win its positions in court. Although SMRF did not get the instream flow water right it requested, one of the founding members, Dr. Jack Fairchild, says that the San Marcos River Foundation is "persistent."**

### Stream Flow

Once water is in the river the fighting has only started. Unlike groundwater, surface water legally belongs to the state, and its use is regulated by official water rights. Surface water from a dependable river like the San Marcos is valuable. It is much cheaper to pump from the river than it is to drill a well. Some towns and cities downstream use the water for municipal supplies. For example, Luling has recently entered a contract to sell river water to Lockhart.

Much of the water withdrawn from the river for municipal use is returned as wastewater. This return flow has traditionally been considered state property once it is put back into the river. However, state law allows cities to "store" that water within the "bed and banks" of the river and to withdraw it downstream. In the 1990s the city of San Marcos applied for a bed and banks permit. The San Marcos River Foundation requested restrictions on the permit that would limit withdrawals in times of low river flow. The city challenged the restrictions imposed by the state, but the courts decided in favor of the San Marcos River Foundation. The Texas Supreme Court refused to consider the city's appeal in 2004, so the entire return wastewater flow may remain in the river, at least for the present.

Why would anyone, especially the San Marcos River Foundation, want

*The San Marcos River is a valuable learning laboratory for the many schools and universities in the region.*

*As the population of Texas continues to increase and become more urban, the natural beauty of all the state's rivers, not only the San Marcos, will become increasingly precious and threatened.*

wastewater left in the river? The reason is that a river is more than water running along a channel. It is a complex habitat, and without sufficient flow the habitat no longer functions. Much of the flow in most Texas rivers, including the San Marcos, consists of treated municipal wastewater that is returned to the river. Without this wastewater, flow in the rivers would diminish drastically. Without sufficient freshwater inflow to the shallow estuaries along the Texas coast, they become too salty to maintain their exceptionally high biological productivity that is fundamental to the commercial and sport fisheries on the coast.

But the question we are now fighting over is how *much* instream and freshwater inflow are necessary? Scientists at the Texas Parks and Wildlife Department (TPWD), which has the major management responsibility for Texas freshwater and coastal fisheries, studied the matter and in 1998 concluded that the San Marcos, along with the Guadalupe that it joins on its way to the coast, should maintain an "environmental" flow of at least 1.15 million acre feet per year at the mouth of the river. Scientists at the Texas Water Development Board agreed with the TPWD estimate. The Texas Water Development Board is the agency responsible for meeting future human water demands, but its officials agreed with the ecologists that the environmental flow is necessary.

In 2000 the San Marcos River Foundation (SMRF) applied to the Texas Natural Resource Conservation Commission (now the Texas Commission on Environmental Quality) for a permit for the 1.15 million acre feet, which they would then donate to the Texas Water Trust, to maintain it in perpetuity for environmental flow. The commission staff supported the application, as did thirty organizations, including commercial fishing and shrimping groups, riverside landowners, recreational fishermen, and environmental groups.

But such a large water allocation means that some other needs or interests may not be served, and strong opposition to the SMRF's application arose. The Guadalupe-Blanco River Authority, Texas Water Conservation Association, the San Antonio River Authority, and the San Antonio Water System vigorously opposed the permit because they did not think a nongovernmental group should have the right to control such a large amount of water.

The fight went to the Texas Legislature, which in 2003 passed legislation that prohibited any allocation of water rights for environmental purposes until September 2005, while a specially appointed committee studied the problem. The legislation did not prohibit any new water rights allocations for other uses, however. The 2005 Texas Legislature did not take action on environmental flows, and

*The San Marcos enters the flood-swollen Guadalupe River just to the left of the center of this photograph. In the future a flood will alter the channels and connect these meandering rivers at some other point.*

Past Creates Future

applications for consumptive uses continue to be filed. The San Marcos River Foundation currently has a lawsuit against the Texas Commission on Environmental Quality regarding denial of its permit application. So rivalries continue on the river.

## Water Quality

The quality of water in the San Marcos River is better than it was thirty years ago, when the polluted water ruined the irrigation system at Ernest Cummings's Green Valley Farm. It is much better than it was sixty years ago when the commander of Gary Field wouldn't let his men swim in the river, or a hundred years ago when William H. Thompson got hepatitis while working on his mill. The water is better because the city of San Marcos spent several million dollars in the 1990s to upgrade its wastewater treatment plant to meet one of the highest standards in the state. Of course, as the city continues to grow it must spend additional money to maintain the quality of its wastewater effluent.

But the current threat to water quality in the San Marcos, as in most rivers in the United States, is not discharges from wastewater treatment systems but runoff from the land. This form of pollution is called nonpoint source, meaning that it does not come from a specific point like the discharge pipe from a wastewater treatment plant. It comes from the entire watershed.

As water runs off the land after a rain, it picks up and carries a wide range of material, depending on how fast the water is running. This material includes particles from tires and brake pads; oil and grease on streets; animal droppings; pesticides and herbicides from lawns and farms; and soil and rocks from fields and construction sites. Solid particles are carried to the river by the physical force of the flowing water. Chemicals are often dissolved in the water.

Federal, state, and local governments all have regulations aimed at preventing or reducing nonpoint source pollution. These regulations attempt to deal with it in two main ways: eliminate the sources and/or catch it before it gets to the river.

Since the San Marcos River receives most of its water from the Edwards Aquifer, which is recharged by rainwater flowing off the Edwards Plateau, all of the contributing watersheds are potential sources of pollution, especially dissolved pollutants. The state's Edwards Aquifer Rules and the city of San Marcos' Edwards Aquifer Protection Regulations require special permits for construction on land that may contribute pollutants to the aquifer. These permits regulate the types and characteristics of development.

All of these laws are in place, so the river is protected, right? Not so. The old saying that "the law is the law" is simplistic. Laws are dynamic. Agencies and courts interpret them, and legislative bodies change them. Environmental laws are especially controversial. While they aim to protect a public good, like water quality, they may impose significant limits on how individuals use their property.

Do you own a nice piece of land in the recharge zone northwest of San Marcos where you would like to build a house beside your pretty little creek? Can't do it. Depending on the size of the creek, the city's Edwards Aquifer Protection Regulations require a water quality zone 50 to 200 feet on each side of it where you cannot build. Outside of the water quality zone is a buffer zone of 100 feet where you are limited to 10 percent impervious cover, unless the slope is 20 percent or greater, in which case you cannot build there. These regulations provide opportunities for mitigation and creative approaches to protecting water quality, but the point is that the city of San Marcos has placed legal limits on the use of land in order to protect the Edwards Aquifer and the San Marcos River.

We Texans place high value on private property rights and do not take such limits lightly. But the aquifer and river are also valuable, and lawmakers have to balance these values. As the saying goes in the environmental protection profession, if everyone is mad at you, you're probably doing a good job!

The upper San Marcos has been greatly changed by materials brought

*The lower river is influenced by local runoff, but due to sparse development in the area, the river remains clean.*

*People express their affection for the San Marcos River in a great variety of ways.*

in by Sessom Creek. A large gravel bar has built up at the mouth of the creek, just below the restaurant near Burleson's dam. The area through the university's Sewell Park has also filled in so that only a narrow channel of fast deep water is left. Analysis of this material shows that it came from construction sites. Some of it was deposited in the river before the city had ordinances that required effective runoff protection from construction sites. Some of the sediment probably came from university projects that are not under the jurisdiction of the city's ordinances.

The city and the university have worked closely with the U.S. Fish and Wildlife Service to develop a Habitat Protection Plan to ensure that the endangered species in the river are protected. Part of the plan involves rerouting Sessom Creek during high flows so that it flows through the old Fish Hatchery ponds, which will catch sediment before it enters the river. Sediment will be removed from the ponds periodically, without affecting the river. The gravel bar at the mouth of the creek will then be removed by careful dredging.

So, water is for fighting. As the population of Texas doubles in the next thirty years and as the Interstate 35 corridor from Georgetown to San Antonio becomes a city of five million people, there will be more people to fight and more things to fight about on the San Marcos River.

*May the water of the beautiful San Marcos River continue to flow downstream to the Guadalupe and the rich estuaries of the Gulf Coast.*

Past Creates Future

Does this mean nothing but a bleak future for the river? Not necessarily. Jacques Cousteau said we take care of what we love. Many, many people love the San Marcos River—the SMRF, the Heritage Society, the "toobers," the birders, the fishers, the Water Safari paddlers, and the rest of us who like to sit on the bank and watch time and the river flow by. The folks in Martindale, Staples, and Fentress love the river enough to live with it. The city of Luling is making plans to celebrate the river in a big way at Zedler's Mill. The Ottine Swamp Monster may be a force to reckon with, not to mention the Texas Commission on Environmental Quality, the Texas Parks and Wildlife Department, the U.S. Army Corps of Engineers, the U.S. Fish and Wildlife Service, and the U.S. Environmental Protection Agency.

What sort of future could this create for the San Marcos River? We could love it to death, with too many people, too many activities. We could pump the aquifer dry and have no river to love. Or we might learn some lessons from the British. The River Thames was the artery of English culture and power, but by the 1860s it was so polluted that Parliament was forced to hang blankets soaked in disinfectant over the windows at Westminster to mask the stench from the river. The invention of the water closet changed the Thames from the main artery of the Empire into its lower intestine, constipated by the daily tides that brought the waste back to the city. By the 1950s the Thames was a dead river. Today salmon have returned to the river. You can catch, and actually eat, them.

What happened? The British people recognized the Thames' importance, spent their money to restore it, and accepted the necessary regulations. The San Marcos is in far better shape than the Thames was and is much smaller. But we must learn how to express our love for the river in a way that protects and enhances it.

The San Marcos River has flowed for millions of years. Our species has used and enjoyed the river for perhaps 12,000 years. But in just 150 years we have changed things that developed naturally over those millions of years. We know how to keep the river flowing and clean. We also know how to dry it up. Which will we choose? The future of the San Marcos River is pretty much our call.

# APPENDIX 1

# Native Species in the Upper San Marcos River

## Aquatic Plants

| Name | Occurence | Characteristics |
|---|---|---|
| *Amblystegium riparium* No common name | common | Moss found on rocks and wood at or near water's edge. Also common in Great Britain. |
| Bladderwort *Utricularia gibba* | occasional | This free-floating plant is carnivorous. Small bladders on the leaves trap and digest tiny animals. |
| Brazilian watermeal *Wolffia papulifera* | common | Smallest flowering plants in the world, leaflike body about 1/16 inch in diameter. Provides food for waterfowl and fish. Species in Africa, India, and Asia are used for cattle and pig feed. A species in Southeast Asia is cultivated and eaten as a vegetable. |
| Broadleaf sagittaria *Sagittaria platyphylla* | common | Waterfowl and beavers eat tubers, which were also used by Indians. |
| Carolina mosquito fern *Azolla caroliniana* | common | This small mosslike plant floats on the water surface, sometimes forming reddish mats. Its leaves contain blue-green algae, which fix nitrogen from the air. Waterfowl feed on the plant, which is used as a fertilizer in Asia. |

| | | |
|---|---|---|
| Coontail<br>*Ceratophyllum demersum* | common | Underwater rootless plant that provides habitat for young fish, aquatic animals, and insects. Not an important food for waterfowl, although some eat the seeds and foliage. Used as an aquarium plant. |
| Creeping primrose willow, Creeping water primrose<br>*Ludwigia repens* | common | Sold as a pond and aquarium plant. Has medicinal uses. |
| Crystalwort<br>*Riccia fluitans* | uncommon | Provides food for waterfowl and other birds and habitat for insects. A liverwort, which was thought to be useful for treating liver diseases in Europe in the Middle Ages. |
| Duckweed (Common)<br>*Lemna minor* | common | Tiny plant with leaflike body about $1/8$ inch in diameter. Provides food for fish and waterfowl. Used as cattle and pig feed in Africa, India, and Asia. |
| Duckweed (Giant)<br>*Spirodela polyrhiza* | common | The leaves of this "giant" plant are about $1/4$ inch in diameter. A high-protein food for ducks, geese, and fish. Used for cattle and pig feed in Africa and Asia. |
| Fanwort<br>*Cabomba caroliniana* | common | Commonly sold as an aquarium plant but is now illegal in some states because its dense growth can be a nuisance. |
| Grassleaf mudplantain or Water stargrass<br>*Heteranthera liebmannii* | occasional | Ducks and other waterfowl eat the leaves. Also provides habitat for fish and insects. |
| Gulf swampweed<br>*Hygrophila lacustris* | uncommon | Harvested and sold for aquaria. |
| Pondweed<br>*Potamogeton illinoensis and nodosus* | common | Provides food and cover for aquatic animals. Tubers of *nodosus* are important food for waterfowl. |

| | | |
|---|---|---|
| Pondweed (Horned) *Zannichellia palustris* | uncommon | Entire plant provides food for waterfowl and other birds and habitat for aquatic animals. |
| Pondweed (Sago) *Potamogeton pectinatus* | uncommon | Seeds, tubers, and vegetation are important food for waterfowl. Diving ducks eat its tubers. |
| Spatterdock *Nuphar luteum* | common | Ducks and other birds eat the seeds. Beaver and nutria eat the roots. Deer may browse the flowers and leaves. Provides spawning habitat for fish. Indians used rhizomes and seeds for food, cooked the roots or ground them into flour. They also ground the seeds into flour or popped them like popcorn. The leaves contain tannin, which is used for dyeing and tanning leather. The leaves were used to stop bleeding and to reduce swelling. |
| Water celery, Tape grass, or Eel grass *Vallisneria americana* | occasional | Diving ducks eat buds on root runners. Other waterfowl and muskrats eat seeds and leaves, which are 1-inch wide and up to several feet long. In Japan people harvest the leaves for food. |
| Watermilfoil (twoleaf) *Myriophyllum heterophyllum* | occasional | Waterfowl eat the fruits. Sometimes sold as an aquarium plant. |
| Water-nymph (Common) *Najas guadalupensis* | occasional | One of the most important food sources for waterfowl. Also provides shelter for insects and small fish. |
| Wildrice (Texas) *Zizania texana* | occasional | This rooted relative of northern and eastern wild rice occurs only in the upper 1.5 miles of the San Marcos River, not including Spring Lake. It is classified as an endangered species. |

## Native Fish

American eel  *Anguilla rostrata*
Bigscale logperch  *Percina macrolepida*
Black buffalo  *Ictiobus niger*
Black bullhead  *Ameiurus melas*
Blackstripe topminnow  *Fundulus notatus*
Blacktail shiner  *Cyprinella venusta*
Blue catfish  *Ictalurus furcatus*
Bluegill  *Lepomis macrochirus*
Bullhead minnow  *Pimephales vigilax*
Central stoneroller  *Campostoma anomalum*
Channel catfish  *Ictalurus punctatus*
Dusky darter  *Percina sciera*
Flathead catfish  *Pylodictus olivaris*
Fountain darter  *Etheostoma fonticola*
Gizzard shad  *Dorosoma cepedianum*
Gray redhorse  *Moxostoma congestum*
Green sunfish  *Lepomis cyanellus*
Guadalupe bass  *Micropterus treculi*
Ironcolor shiner  *Notropis chalybaeus*
Largemouth bass  *Mircopterus salmoides*
Largespring gambusia  *Gambusia geiseri*
Longear sunfish  *Lepomis megalotis*
Longnose gar  *Lepisosteus osseus*
Mimic shiner  *Notropis volucellus*
Orangethroat darter  *Etheostoma spectabile*
Redear sunfish  *Lepomis microlophus*
Red shiner  *Cyprinella lutrensis*
River carpsucker  *Carpiodes carpio*
Roundnose minnow  *Dionda episcopa*
Smallmouth buffalo  *Ictiobus bubalus*
Speckled chub  *Macrhybopsis aestivalis*
Spotted bass  *Micropterus punctulatus*
Spotted gar  *Lepisosteus oculatus*
Spotted sunfish  *Lepomis punctatus*
Texas logperch  *Percina carbonaria*
Texas shiner  *Notropis amabilis*
Warmouth  *Lepomis gulosus*
Western mosquitofish  *Gambusia affinis*
White crappie  *Pomoxis annularis*
Yellow bullhead  *Ameiurus natalis*

# APPENDIX 2

# Nonnative Species in the Upper San Marcos River

## Aquatic Plants

| Name | Status | Characteristics |
|---|---|---|
| Brazilian elodea or leafy elodea *Egeria densa* | established | Introduced from South America. Sold as an aquarium oxygenator plant under the name Anacharis. Displaces native plants and forms dense beds that interfere with water flow. |
| Cryptocoryne *Cryptocoryne beckettii* | established | Introduced from Southeast Asia as an aquarium plant. Currently found only near the outlet of the San Marcos Wastewater Treatment Plant. |
| Cursed buttercup *Ranunculus sceleratus* | established | Introduced from Europe as an ornamental. It is likely "cursed" because its sap can cause skin blisters. Found around the edges of Spring Lake. |
| Elephant ear *Colocasia esculenta* | established | Introduced to the San Marcos River in the early twentieth century. Originally brought to North America from Southeast Asia as an ornamental. Dominates the banks of the river past the confluence with the Blanco. Causes substantial losses of water from the river through its high rate of evapotranspiration. |

| | | |
|---|---|---|
| Elephant ear *Xanthosoma sagittifolium* | established | Native to the West Indies and tropical South America. Used as an ornamental. Found in the San Marcos River only below the spillways of Spring Lake and Rio Vista dams. |
| Floating fern *Ceratopteris thalictroides* | established | Native of East Asia and Australia. Cultivated as an ornamental in England in the 1800s and currently grown in Japan as a vegetable. Sold for use in aquaria and ponds. Introduced in the San Marcos River by the Texas Aquatic Plants Company. |
| Giant reed *Arundo donax* | established | Introduced from the Mediterranean, probably during the 1800s. Found along the banks and on sandbars. Clumps of the reed trap materials flowing downstream and form islands that alter the channel. |
| Hydrilla *Hydrilla verticillata* | established | Native to Africa, Australia, and Asia. The most problematic aquatic plant in the U.S. due to its dense growth and ability to grow in many conditions. |
| Indian hygrophila *Hygrophila polysperma* | established | Popular aquarium plant introduced from Malaysia and India. It has been in the San Marcos River for at least twenty-five years and has become a dominant species, taking habitat from native species. |
| Limnophila *Limnophila sessiliflora* | established | Native to India, Ceylon, and the Philippines. Originally brought to U.S. and sold as aquarium plant. Stems can grow up to 12 feet long. Regrows from fragments. Can shade out other species. |
| Pondweed (Curly) *Potamogeton crispus* | established? | Aquarium plant introduced from Europe. It seemed to be well established in the mid-1980s, but recent researchers have not found it. |
| Watercress *Rorippa nasturtium-aquaticum* | established | This edible European plant was introduced to the river by European settlers. |

| | | |
|---|---|---|
| Water hyacinth<br>*Eichhornia crassipes* | established | The purple bloom made this South American plant popular for ornamental ponds. The Japanese delegation gave these plants to participants at a cotton exposition in New Orleans in 1884. They have been common in the San Marcos River since at least the early 1920s. Considered to be one of the world's ten worst invasive plants. Populations in the San Marcos River are small and found mainly along the edges of Spring Lake and backwaters of the river. Ecosystem restoration activities at Spring Lake now include removal of water hyacinth. |
| Water lettuce<br>*Pistia stratioles* | established | Botanists disagree whether this plant is native to the U.S., but they agree that it is not native to central Texas. It was probably introduced to the San Marcos River from East or South Texas, sometime before the early 1920s. It has been found only in Spring Lake and backwaters of the river. |
| Watermilfoil (Eurasian)<br>*Myriophyllum spicatum* | established | Native to Eurasia and North Africa. Extremely fast-spreading plant that can reproduce from fragments carried by boats. May have spread as packing material for worms sold to fishermen. Considered to be one of the worst invasive aquatic weeds, it crowds out native species and blocks sunlight and oxygen. Botanists have not observed the plant in the San Marcos River in recent years. |
| Watermilfoil (Parrot feather)<br>*Myriophyllum aquaticum* | established | Introduced from South America as an aquarium plant. Creates dense mats on the surface. Very tolerant to droughts and freezes. However, parrot feather occurs only in small populations in Spring Lake and backwaters of the river. |

## Nonnative Fish

Amazon molly   *Poecilia formosa*
Blue tilapia   *Oreochromis aureus*
Common carp   *Cyprinus carpio*
Convict cichlid   *Cichlasoma nigrofasciatum*
Goldfish   *Carrasius auratus*
Inland silversides   *Menidia beryllina*
Mexican tetra   *Astyanax mexicanus*
Mossambique tilapia   *Oreochromis mossambicus*
Pacu   *Colosomma sp.*
Redbreast sunfish   *Lepomis auritus*
Rift-lake cichlid   *Pseudotropheus sp.*
Rio Grande cichlid   *Cichlasoma cyanoguttatum*
Rock bass   *Amblopities rupestris*
Sailfin molly   *Poecilia latipinna*
Smallmouth bass   *Micropterus dolomieu*
Suckermouth catfish   *Hypostomus sp.*
Vermiculated sailfin catfish   *Pterygoplichthys disjunctivus*

# APPENDIX 3

# Summary of Edwards Aquifer Habitat Conservation Plan Measures

Edwards Recharge and/or recirculation enhancement

Precipitation enhancement

Water quality protection

Water conservation and reuse requirements

Implementation of alternative management practices, procedures, or methods allowed by the EAA

Pumping withdrawals will be determined by initial and additional regular permits above aquifer level 665' as measured by Index Well J-17 and above 865' as measured by Index Well J-27. Aquifer withdrawals will be reduced to 449,950 acre-feet per year when aquifer levels fall below 665' for J-17 and 865' for J-27.

When the aquifer level declines to 665', total pumpage will be 449,950 ac-ft/yr. Annual water budget required for each pumper, with four stage DM/CPM reductions when the aquifer reaches the following levels: Stage 1 (J-17 = 650': maximum pumpage = 436,300 ac-ft/yr); Stage II (J-17 = 640': maximum pumpage = 422,800 ac-ft/yr).; Stage III (J-17 = 630' or J-27 = 845': maximum pumpage = 382,000 ac-ft/yr); Stage IV (J-17 = 627' or J-27 = 842': maximum pumpage = 346,400 ac-ft/yr).

Assist with funding field collection and distribution of species to refugia.

Assist with funding new salamander facility at the San Marcos National Fish Hatchery.

Assist with funding refugia for existing captive stock at San Marcos National Fish Hatchery.

Assist with funding refugia for existing captive stock at Uvalde National Fish Hatchery.

Assist with funding salvage of additional species for refugia.

Fund costs for personnel labor to manage and maintain refugia.

Aquatic vegetation enhancement (reintroduction/establishment of native species) in select areas.

Aquatic vegetation restoration (reintroduction/reestablishment of native species) after low-flow events.

Continued evaluation of aquatic vegetation responses to low flow/elevated temperature.

Management/research to determine parasite impact to fountain darter (current EAA Variable Flow Study).

Improve accuracy of USGS gauges below Spring Lake and Landa Lake (ongoing).

Establish discharge monitoring gauge on Old Channel of Comal River.

- Continue evaluation of drought survival mechanisms of the Comal Springs riffle beetle; low-flow laboratory evaluations, and subsequent field-based study of hyporheic population density (preliminary study completed).
- Determine life history requirements of the three endangered invertebrates, including population dynamics, distribution, tolerance/sensitivity (temperature, water quality, contaminants), and reproduction.
- Establish water quality monitoring network of three wells near Comal and San Marcos Springs.
- Develop and implement management plan for vegetation mat removal during low flow.
- Refine estimate of amount of pumpage from exempt wells.
- Determine effects of contaminants on Covered Species.
- Determine gains and losses to instream flows in the Guadalupe River.
- Water Quality/Variable Flow Monitoring Study (ongoing).
- Studies to determine tolerance of individual species to the ranges of various water quality parameters expected with on-site intensive management areas.
- Pilot study of intensive management areas in both the San Marcos and Comal Rivers.

# REFERENCES

*Chapter 1: More than Just a Little River*

Amos, Bonnie B., and Frederick R. Gehlbach. 1988. *Edwards Plateau Vegetation: Plant Ecological Studies in Central Texas.* Waco: Baylor University Press.

Bomar, George. 1983. *Texas Weather.* Austin: University of Texas Press.

Bonnell, Geo. W. 1840. *Topographical Description of Texas. To Which is Added, an Account of the Indian Tribes.* Austin: Clark, Wing, and Brown. Repr., Waco, Tex.: Texian Press, 1964.

Brune, Gunnar. 1975. Report 189. *Major and Historical Springs of Texas.* Austin: Texas Water Development Board.

*Daily Valley Morning Star.* 1970. San Marcos Flood Liberates Alligators. May 17.

Earl, Richard A. and Charles R. Wood. 2002. Upstream Changes and Downstream Effects of the San Marcos River of Central Texas. *Texas Journal of Science* 54 (1): 69–88.

Eckhardt, Gregg. Edwards Aquifer Homepage. San Marcos Springs. http://www.edwardsaquifer.net/sanmarcos.html. (accessed January 22, 2001).

Harrigan, Stephen. 1994. *A Natural State: Essays on Texas.* Austin: University of Texas Press.

*Hays County Times.* October 2, 1895.

Huser, Verne. 2000. *Rivers of Texas.* College Station: Texas A&M University Press.

*Jacksonville Daily Progress.* May 17, 1970. San Marcos Flood Worst Ever to Hit College Town.

Jordan, Terry G., with John L. Bean Jr. and William M. Holmes. 1984. *Texas: A Geography.* Boulder, Colo.: Westview Press.

Marcee, Clarisa. 1999. Sacred Springs. *Texas Parks and Wildlife* 57 (6): 22A–25A.

*San Antonio Express.* 1970. Last of Aquarena Alligators Found. May 26.

*San Marcos Daily Record.* 1974. Flooding Hits San Marcos Area. November 24.

Saunders, Kenneth S., Kevin B. Mayes, Tim A. Jurgensen, Joseph F. Trungale, Leroy J. Kleinsasser, Karim Aziz, Jacqueline Renee Fields, and Randall E. Moss. 2001. *An Evaluation of Spring Flows to Support the Upper San Marcos River Spring Ecosystem, Hays County, Texas.* Austin: Resource Protection Division, Texas Parks and Wildlife Department.

Slade, R. M., and Kristie Persky. 1999. *Floods in the Guadalupe and San Antonio River Basins in Texas, October 1998.* Austin: U.S. Geological Survey.

Soil Conservation Service. 1978. *Watershed Plan and Environmental Impact Statement: Upper San Marcos River Watershed.* Temple, Tex.: U.S. Department of Agriculture.

Spearing, Darwin. 1991. *Roadside Geology of Texas.* Missoula, Mont.: Mountain Press Publishing Company.

Stephens, A. Ray, and William M. Holmes. 1989. *Historical Atlas of Texas.* Norman: University of Oklahoma Press.

Swanson, Eric R. 1995. *Geo-Texas: A Guide to the Earth Sciences.* College Station: Texas A&M University Press.

U.S. Army Corps of Engineers. 1971. *Flood Hazard Information, San Marcos and Blanco Rivers, San Marcos, Texas.* Fort Worth, Tex.: U.S. Army Corps of Engineers.

U.S. Geological Survey. Peak Streamflow. http://nwis.waterdata.usgs.gov/nwis/peak/?site_no=08172000 (accessed October 15, 2002).

U.S. Geological Survey. Surface Water Data for Texas: Calendar Year Streamflow Statistics. http://waterdata.usgs.gov/tx/nwis/annual/calendar (accessed October 15, 2002).

Wiese, Art. 1972. Death Toll 11 in Central Texas Flood. *Houston Post.* May 13.

Woodruff, C. M., Jr. 1977. Stream Piracy near the Balcones Fault Zone, Central Texas. *Journal of Geology* 85 (4): 483–90.

Woodruff, C. M., Jr., and P. L. Abbott. 1979. Drainage-basin Evolution and Aquifer Development in a Karstic Limestone Terrain, South-Central Texas, USA. *Earth Surface Processes* 4 (4): 319–34.

## Chapter 2: Life of the River

Bartlett, Richard C. 1995. *Saving the Best of Texas: A Partnership Approach to Conservation.* Austin: University of Texas Press.

Bowles, D. E., and B. D. Bowles. 2001. *A Review of the Exotic Species Inhabiting the Upper San Marcos River, Texas, U.S.A.* Austin: Texas Parks and Wildlife Department.

Bowles, David E., Karim Aziz, and Charles L. Knight. 2000. *Machrobrachium* (Decapoda: Caridea: Palamemonidae) in the United States: A Review of the Species and an Assessment of Threats to Their Survival. *Journal of Crustacean Biology* 20 (1): 158–71.

Boxall, Bettina. 1976. San Marcos River Plants Also Used in Aquariums. *San Marcos Daily Record.* June 18.

Brazoria County Historical Museum. Texas Agricultural Experiment Station, Angleton, Texas. http://www.bchm.org (accessed December 11, 2002).

City of San Marcos. 1985. Environmental Assessment of the San Marcos River.

Coley, Ron. 2004. Interview by the author. June 16.

Doughty, Robin, and Barbara M. Parmenter. 1989. *Endangered Species: Disappearing Animals and Plants in the Lone Star State.* Austin: Texas Monthly Press.

Devall, Lonnie L. 1940. A Comparative Study of Plant Dominance in a Spring-fed Lake. Master's thesis, Southwest Texas State Teachers College.

Elphick, Chris, John B. Dunning Jr., and David Allen Sibley. 2001. *The Sibley Guide to Bird Life and Behavior.* New York: Alfred A. Knopf.

Floridata Marketplace. *Sabal minor.* http://www.floridata.com/ref/s/saba_min.cfm.

Garrett, Judith M., and David G. Barker. 1987. *A Field Guide to Reptiles and Amphibians of Texas.* Houston: Gulf Publishing Company.

Gee, Robert W. 2000. Rice Plants Shrivel After Dam Breach. *Austin American Statesman,* January 4.

Georgia Wildlife Web Site. Mammals: *Sylvilagus aquaticus.* http://museum.nhm.uga.edu/gawildlife/mammals/lagomorpha/leporidae/saquaticus.html (accessed December 19, 2002).

Griffith, G. E., S. A. Bryce, J. M. Omernik, J. A. Comstock, A. C. Rogers, B. Harrison, S. L. Hatch, and D. Bezanson. 2004. *Ecoregions of Texas* (color poster with map, descriptive text, and photographs). Reston, Va.: U.S. Geological Survey.

Gulf States Marine Fisheries Commission. *Macrobrachium olfersii* (Wiegmann, 1836). http://nis.gsmfc.org/nis_factsheet.php?toc_id=142.

Lemke, David E. 1989. Aquatic Marcophytes of the Upper San Marcos River, Hays Co., Texas. *Southwestern Naturalist* 34:289–91.

*The Mammals of Texas* online edition. Nutria. http://www.nsrl.ttu.edu/tmot1/myoccoyp.htm.

McMahan, Craig A., Roy G. Frye, and Kirby L. Brown. 1984. *The Vegetation Types of Texas Including Cropland.* Austin: Texas Parks and Wildlife Department.

Niering, William A. 1985. *Wetlands.* New York: Chanticleer Press Inc.

Paqtnkek Fish and Wildlife Commission. Kat (American Eel). http://www.stfx.ca/research/srsf/Fact%20Sheets/FS6.htm.

*The Potato of the Humid Tropics.* http://www.botgard.ucla.edu/html/botanytextbooks/economicbotany/Colocasia/.

*San Marcos Daily Record.* 1962. Underwater Farm Unique.

Texas Parks and Wildlife Department. Nature Checklist of Texas Wildlife. Scientific and Common Names of Texas Waterfowl. http://www.tpwd.state.tx.us/huntwild/wild/species/

Thomas, Alexander. 1992. *River of Innocence, River of Progress: A Brief Environmental History of the San Marcos River.* San Marcos: Southwest Texas State University.

University of Michigan Museum of Zoology. *Anguilla rostrata.* http://animaldiversity.ummz.umich.edu/accounts/anguilla/a._rostrata$narrative.html.

Whiteside, Bobby G., Alan W. Groeger, Patrick F. Brown, and Travis C. Kelsey. 1994. Physiochemical and Fish Survey in Spain, Robert W. (project coordinator). *The San Marcos River: A Case Study.* Austin: Texas Parks and Wildlife Department.

Williamson, Leroy. 1990. River on the Edge. *Texas Parks and Wildlife* 48 (12): 18–23.

## Chapter 3: From the Deep Past at the Springs

Barrera, Jimmy. 2002. *San Marcos City Park Archaeological Survey, Hays County, Texas.* Archaeological Studies Report No. 4. San Marcos: Center for Archaeological Studies, Southwest Texas State University.

Baumgartner, Dorcas, William C. Foster, and Jack Jackson. 1997. *Frontier River: Exploration and Settlement of the Colorado River.* Austin: Lower Colorado River Authority.

Bousman, C. Britt. 1998. Paleoenvironmental Change in Central Texas: The Palynological Evidence. *Plains Anthropologist* 43 (164): 201–19.

———. n.d. The Texas River Center Archaeology Project: Phase 1 Testing. San Marcos: Center for Archaeological Studies, Southwest Texas State University.

Bousman, C. Britt, David L. Nickels, Kevin Schubert, Barbara Meissner, Isabel Gutierrez Sr., and Vergie Richardson. 2003. *Archaeological Testing of the Burleson Homestead at 41HY37 Hays County, Texas.* San Marcos: Center for Archaeological Studies, Southwest Texas State University.

Bryant, Vaughn M. Paleoenvironments. *The Handbook of Texas Online.* http://www.tsha.utexas.edu/handbook/online/articles/PP/sop2.html.

Cabeza de Vaca, Alvar Núñez. 2002. *Chronicle of the Narváez Expedition.* Revised and annotated by Harold Augenbraum, with an introduction by Ilan Stavans. New York: Penguin Books. (Orig. pub. 1542. Trans. by Fannie Bandelier, 1905.)

California Indian Acorn Culture. http://archives.gov/pacific/education/4th-grade/acorn.html.

Castaneda, Carlos E. 1958. *Our Catholic Heritage in Texas, 1519–1936.* Austin: Von Boeckmann-Jones. (Orig. pub. 1936.)

Chipman, Donald E. 1992. *Spanish Texas, 1519–1821.* Austin: University of Texas Press.

De la Teja, Jesús F., Paula Marks, and Ron Tyler. 2004. *Texas: Crossroads of North America.* Boston: Houghton Mifflin Company.

Dobie, J. Frank, ed. 1975. *Lost Mines and Buried Treasure.* Vol. 1 of *Legends of Texas.* Gretna, La.: Pelican Publishing Company. (Orig. pub. 1924.)

Fehrenbach, T. R. *Lone Star: A History of Texas and the Texans.* 1991. New York: Wings Books. (Orig. pub. 1968.)

Foster, Nancy Haston. 1995. *Texas Missions.* Houston: Gulf Publishing Company.

Foster, William C. 1995. *Spanish Expeditions into Texas 1689–1768.* Austin: University of Texas Press.

Garber, James F. 1985. Possible Transitional Archaic House Structure and Activity Areas at 41HY163 San Marcos, Texas. Presented at the annual meeting of the Texas Archaeological Society in San Antonio, Texas, November 2.

Hatcher, Mattie Austin, trans. 1932. "The Expedition of Don Domingo Terán de los Ríos into Texas." *Preliminary Studies of the Texas Catholic Historical Society* 2 (1): 10–67.

Hester, Thomas R. 1986. Early Human Populations along the Balcones Escarpment. In *The Balcones Escarpment, Central Texas,* edited by Patrick Abbott and C. M. Woodruff, 55–62. San Antonio: Geological Society of America.

Hester, Thomas R., and Ellen Sue Turner. Prehistory. *The Handbook of Texas Online.* http://www.tsha.utexas.edu/handbook/online/articles/view/PP/sop2.html.

Illinois State Museum. Mammoths. http://www.museum.state.il.us/exhibits/larson/mammuthus.html.

John, Elizabeth A. H. 1975. *Storms Brewed in Other Men's Worlds: The Confrontation of Indians, Spanish, and French in the Southwest, 1540–1795.* Lincoln: University of Nebraska Press.

Jones, Richard S. 2002. *Archaeological Survey and Testing of Schulle Park, Hays County, Texas.* Archaeological Studies Report No. 2. San Marcos: Center for Archaeological Studies, Southwest Texas State University.

Keddie, Grant. The Atlatl Weapon. Royal British Columbia Museum. http://www.royalbcmuseum.bc.ca.

Lundelius, Ernest. 1986. Vertebrate Paleontology of the Balcones Fault Trend. In *The Balcones Escarpment, Central Texas,* edited by Patrick Abbott and C. M. Woodruff, 41–50. San Antonio: Geological Society of America.

McGraw, A. Joachim, John W. Clark Jr., and Elizabeth A. Robbins, eds. 1998. *A Texas Legacy: The Old San Antonio Road and The Caminos Reales: A Tricentennial History, 1691–1991.* Austin: Texas Department of Transportation.

Nickels, David L., and C. Britt Bousman. 2002. *An Archaeological and Geoarchaeological Investigation of 5.15 Acres for the San Marcos Consolidated Independent School District, Hays County, Texas.* Technical Report 1. San Marcos: Center for Archaeological Studies, Southwest Texas State University.

Nordt, Lee, and Britt Bousman. 2001. Preliminary Geoarchaeology of Aquarean Springs. Unpublished Report.

Prieto, Carlos. 1973. *Mining in the New World.* New York: McGraw-Hill Book Company.

Shiner, Joel L. 1983. Large Springs and Early American Indians. *Plains Anthropologist* 28 (99): 1–6.

Takac, Paul R. "Homebases" and the Paleoindian/Archaic Transition in Central Texas. Unpublished reading copy.

University of Texas at Austin, College of Liberal Arts. Texas Beyond History. Graham-Applegate. http://www.texasbeyondhistory.net/graham/austin.html.

Weddle, Robert S. 1968. *San Juan Bautista: Gateway to Spanish Texas.* Austin: University of Texas Press.

## Chapter 4: Anglo Americans at the Springs

Hunter, J. Marvin. 1985. *The Trail Drivers of Texas.* Austin: University of Texas Press. (Orig. pub. 1924.)

Jackson, Norris. 1952. Mermaid Theater. *Popular Mechanics.* June.

Jordan, Terry G. 1993. *North American Cattle-Ranching Frontiers: Origins, Diffusion, and Differentiation.* Albuquerque: University of New Mexico Press.

Kilstofte, June. 1950. Over the Devil's Backbone. *San Antonio Express Magazine.* June.

Lehman, Shirley Rogers. 2002. Interview by the author. November 15.

McClintock, William A. Journal of a Trip Through Texas and Northern Mexico in 1846–1847. *Southwestern Historical Quarterly* 34 (1): 32–33.

McConnell, H. H. 1996. *Five Years a Cavalryman; or Sketches of Regular Army Life on the Texas Frontier, 1866–1871.* Norman: University of Oklahoma Press. (Orig. pub. 1889.)

McGehee, Albert S. n.d. *A River Reflects on Pepper's Past.* Privately published pamphlet.

———. 1989. *The Inn at the Head of the River.* Privately published pamphlet.

McGehee, Scott. 2002. Interview by the author. November 20.

Meinig, D. W. 1969. *Imperial Texas: An Interpretive Essay in Cultural*

*Geography.* Austin: University of Texas Press.

Muir, Andrew Forest, ed. 1958. *Texas in 1837: An Anonymous, Contemporary Narrative.* Austin: University of Texas Press.

Roemer, Ferdinand. 1935. *Texas with Particular Reference to German Immigration and the Physical Appearance of the Country.* Translated from German by Oswald Mueller. San Antonio: Standard Printing Company.

*San Augustine (FL) Record.* 1950. Marine Studios to Build Underwater Theatre in Texas. May 18.

*San Marcos Record.* 1949. Rogers Begins Preliminary Work Dredging for Underwater Theater. November 18.

*San Marcos Record.* 1950. Construction of Underwater Theater to Begin Within Week. May 23.

*San Marcos Record.* 1950. Construction of Aquarena 60 Percent Complete. July 14.

*San Marcos Record.* 1952. Bountiful Water Supply Sets San Marcos Apart in Texas in Times of Widespread Scarcity. August 29.

*San Marcos Record.* 1955. Aquarena Launches Expansion; Expects Record Crowd in 1956. December 16.

*San Marcos Record.* 1958. Century-Old House to Be Moved, Restored. January 8.

*San Marcos Record.* 1960. Aquarena Plans $300,000 Additions With Entrance, Fountain, Skyride. November 14.

*San Marcos Record.* 1963. New Sky-Ride Crossing Spring Lake. May 23.

*San Marcos Record.* 1965. "Tarzan" Weissmuller Will Visit Aquarena for Television Series. March 18.

*San Marcos Record.* 1965. Rites Sunday For Founder of Aquarena, Paul J. Rogers. September 23.

*San Marcos Record.* 1967. Aquarena Slates Volcano. March 25.

*San Marcos Record.* 1972. Early San Marcos History Unveiled in Letter. August 17.

*San Marcos Record.* 1976. Aquarena Volcano Burns. February 22.

*San Marcos Record.* 1978. Piranha: It's Movie Time Again Here. March 22.

*San Marcos Record.* 1981. Restaurant Plans Displace Swimmers. April 21.

Shelton, Ed. 1947. Underwater Gardens. *San Antonio Express Magazine.* November 2.

Tolbert, Frank X. 1964. Above the River, A Hanging Garden. *Dallas Morning News.* June 13.

Underwood, Bibb. 2000. Aquarena Springs: A Look Back at the Beginning . . . *San Marcos Daily Record.* October 29.

Ward, Robin Stacey. 2001. Unpublished manuscript about Aquarena Springs.

Williamson, Hugh. 1962. How To Attract Tourists by Trying Hard for a Long Time. *Texas Parade.* April.

## Chapter 5: San Marcos the River, San Marcos the Town

Boxall, Bettina. 1976. San Marcos River Plants Also Used in Aquariums. *San Marcos Daily Record.* June 18.

Connally, Preston, n.d. Unpublished autobiography.

Connally, Preston and Doris. 2002. Interview by the author. June.

Cornell, Paul. 2001. Interview by the author. July 18.

Davis, Steve, ed. 2000. *Sueños y recuerdos del pasado: A Community History of Mexican Americans in San Marcos.* San Marcos: Hispanic Historical Committee, Hays County Historical Commission.

Hildebrandt, Gene. 1982. Rio Vista: Old Dam's History Winds through the Courts. *San Marcos Daily Record.* October 6.

Leimer, Christina. 1994. Down by the Old Mill Stream. *Texas Parks and Wildlife Magazine.* June.

McDonald, Vernon. 2001. Interview by the author. July 19.

Moore, Billy. 2004. Interview by the author. June 3.

Rich, Kathryn Thompson. 2004. Interview by the author. June 24.

Rich, Kathryn Thompson, and Tula Townsend Wyatt. 1978. *Thompson's Islands.* Unpublished manuscript.

Sanborn, Chloe (Walker). 1944. *The Story of Riverside.* Master's thesis, Southwest Texas State Teachers College.

*San Marcos Daily Record.* 1962. Underwater Farm Unique. (Tula Townsend Watt file, San Marcos Public Library.)

*San Marcos Daily Record.* 1973. 1840 Post San Marcos Recognized. September 16.

Smith, Gwen. 2004. Interview by author (May 21), and personal records.

Stovall, Frances, Maxine Storm, Louise Simon, Gene Johnson, Dorothy Schwartz, and Dorothy Wimberley Kerbow. 1986. *Clear Springs and Limestone Ledges, A History of San Marcos and Hays County for the Texas Sesquicentennial.* Austin: Nortex Press.

Taylor, T. U. 1904. *The Water Powers of Texas.* Washington, D.C.: U.S. Geological Survey.

Wimberley, C. W. 1965. *Stone Milling and Whole Grain Cooking.* Austin: Von Boeckmann-Jones Press.

———. 1969. *Wimberley Hills: A Pioneer Heritage.* Austin: Von Boeckmann-Jones Press.

———. 1989. My River of Innocence, parts 2, 3, and 4. *San Marcos Daily Record.* September.

## Chapter 6: Dams and Towns: A Nostalgic River Landscape

Baker, T. Lindsay. 1986. *Building the Lone Star: An Illustrated Guide to Historic Sites.* College Station: Texas A&M University Press.

Britton, Karen Gerhardt, Fred C. Elliott, and E. A. Miller. Cotton Culture. *The Handbook of Texas Online.* http://www.tsha.utexas.edu/handbook/online/articles/view/CC/afc3.html (accessed August 26, 2002).

Cummings, Ernest. 2004. Interview by the author. May 28.

De la Teja, Jesús Francisco, Paula Marks, and Ron Tyler. 2004. *Texas: Crossroads of North America.* Boston: Houghton Mifflin Company.

Dobie, J. Frank. 1974. *Guide to Life and Literature of the Southwest.* Dallas: Southern Methodist University Press. (Orig. pub. 1942.)

Fehrenbach, T. R. 1968. *Lone Star: A History of Texas and the Texans.* New York: Wings Books.

Frantz, Joe B., and Mike Cox. 1988. *Lure of the Land: Texas County Maps and the History of Settlement.* College Station: Texas A&M University Press.

Hardin, Stephen L. Ottine, Texas. *The Handbook of Texas Online.* http://www.tsha.utexas.edu/handbook/online/articles/oo/hno21.html (accessed February 15, 2001).

———. Slayden, Texas. *The Handbook of Texas Online.* http://www.tsha.utexas.edu/handbook/online/articles/SS/hns52.html (accessed February 13, 2001).

Hilderbrandt, Gene. 1982. Rio Vista: Old Dam's History Unwinds Through the Courts. *San Marcos Daily Record.* October 6.

Holmes, Martha Nell. 2004. Interview by the author. June 18.

Kownslar, Allan O. 2004. *The European Texans.* College Station: Texas A&M University Press.

Lanning, James W. 2003. *Life & Memories of Ernest E. Cummings: Growing Up With the Pipeline Industry.* San Marcos: James W. Lanning.

Lowman, Al. 2004. Interview by the author. June 16.

Lowman, Al. Staples, Texas. *The Handbook of Texas Online.* http://www.tsha.utexas.edu/handbook/online/articles/SS/hns78.html (accessed January 28 2001).

Netardus, Leon. n.d. *Ghosts of Gonzales.* Gonzales, Tex.: Reese's Print Shop.

Rundell, Walter, Jr. 1977. *Early Texas Oil: A Photographic History, 1866–1936.* College Station: Texas A&M University Press.

Smyrl, Vivian Elizabeth. Luling, Texas. *The Handbook of Texas Online.* http://www.tsha.utexas.edu/handbook/online/articles/LL/hj117.html (accessed January 28, 2001).

———. Stairtown, Texas. *The Handbook of Texas Online.* http://www.tsha.utexas.edu/handbook/online/articles/SS/hnsve.html (accessed February 9, 2004).

Stock, Barbara. Fentress, Texas. *The Handbook of Texas Online.* http://www.tsha.utexas.edu/handbook/online/articles/FF/hnf16.html (accessed January 28, 2001).

———. Prairie Lea, Texas. *The Handbook of Texas Online.* http://

www.tsha.utexas.edu/handbook/online/articles/PP/hlp50.html (accessed January 28, 2001).

Syers, Ed. 1981. The Thing in Ottine Swamp. In *Ghost Stories of Texas*, 22–25. Waco: Texian Press.

Taylor, T. U. 1904. *The Water Powers of Texas*. Washington, D.C.: U.S. Geological Survey.

Texas State Data Center and Office of the State Demographer. 2003 Total Population Estimates for Texas Places. http://txsdc.utsa.edu/tpepp/2003_txpopest_place.php.

Valenza, Janet Mace. 2000. *Taking the Waters in Texas: Springs, Spas, and Fountains of Youth*. Austin: University of Texas Press.

Wagner, Scott E. Martindale, Texas. *The Handbook of Texas Online*. http://www.tsha.utexas.edu/handbook/online/articles/MM/hlm32.html (accessed January 28, 2001).

Wyatt, Tula Townsend. Camp Clark. *The Handbook of Texas Online*. http://www.tsha.utexas.edu/handbook/online/articles/CC/qcc7.html (accessed January 28, 2001).

Zedler, Donald. 1996. *The Diary of Sofie Marie Zedler*. Kearney, Neb.: Morris Publishing.

## Chapter 7: Past Creates Future

Berryhill, Michael. 2003. San Antonio Bay: The Whooper's Table. *Texas Parks and Wildlife* 61 (7): 38–48.

Blackburn, Jim. 2004. *The Book of Texas Bays*. College Station: Texas A&M University Press.

Britton, Joseph C., and Brian Morton. 1989. *Shore Ecology of the Gulf of Mexico*. Austin: University of Texas Press.

Edwards Aquifer Authority. 2004. Evaluation of Augmentation Methodologies in Support of In-situ Refugia at Comal and San Marcos Springs, Texas. LBG-Guyton Associates, in association with BIO-WEST, Inc., Espey Consultants, Inc., and URS Corporation.

Edwards Aquifer Authority. 2004. Report to House Natural Resources Committee. September 20.

Edwards Aquifer Authority. 2005. Draft Edwards Aquifer Habitat Conservation Plan. Prepared by Hicks & Company/RECON in association with Bio-West, Inc., LBG-Guyton Associates.

Longley, Glenn. 2004. Interview by the author. June 15.

Moss, Randy. 2004. Interview by the author. June 18.

Sayre, Nathan F. 2001. *The New Ranch Handbook: A Guide to Restoring Western Rangelands*. Santa Fe, N.M.: The Quivira Coalition.

Smeins, F. E., and S. D. Fuhlendorf. Biology and Ecology of the Ashe (Blueberry) Juniper. http://texnat.tamu.edu/symp/juniper/FRED2.htm (accessed September 12, 2000).

Wassenich, Dianne, executive director, San Marcos River Foundation. 2002, 2005. Interviews by the author. June 18. August 31.

# INDEX

*Italicized numbers refer to illustration captions.*

Abbott, Patrick, 5
acidic water, 4, 10
acorns, 33, 34–36
agricultural chemicals, 10
Alabama, state of, 106
Alcot turbine, 106, 108, 111
alligators, 23, 33, 36, 52
alluvial soil, 91, 94
American eel, 21
American Museum of Natural History, 61
Anglo Americans, 3, 10, 47, 49, 68, 94, 106
*A Perfect World*, 100
Aquamaids, 61, 98
Aquarena Center, 16, 65
Aquarena Springs, 6, 50, 57, 61, 64, 64–65, 71, 86, 89, 98
archaeologists, 34, 35, 36, 38
Arizona, state of, 45
Army Air Corps, 59
Asian clam, 28
Atlantic Ocean, 21
Atlas steam engine, 70, 102–103
atlatl, 33, 36, 38
Austin, Moses, 52
Austin, Stephen F., 52, 94
Austin, Texas, 3, 4, 18, 49, 51–52, 62, 68, 70–72, 101, 103

Bachman, J. A., 96
backyard privies, 76
badger, 33
Baker, T. Lindsay, 115
Balcones Canyonlands, 10
Balcones Escarpment and Fault, 3, 4, 23, 42
Bamberger, J. David, 122
Bamberger Ranch Preserve, 122
barge, 59
base flow, 7
bass, 15
Bastrop, Texas, 49, 51, 54
bathhouse, 69, 85, 109
bathing regatta, 78
Baugh, John, 63
bears, 36, 40, 71
Beauty Along the River, 81
bed and banks, 122
Beidler, Ed, 70
Bellinger, Edmund, 105
Berryhill, Michael, 120
big claw river shrimp, 21
birds, 30; black swans, 28; European mute swan, 28, 30; great blue herons, 15, 23, 24; kingfishers, 15; red-shouldered hawk, 15, 23; turkeys, 36, 40; whooping cranes, 120
Bishop Doggett, 79
bison, 33, 34
blackberries, 36
Blackland Prairie, 10, 11, 12, 16, 92, 117
Black Pipe Bridge, 57, 79, 80, 81
Blanco River, 2, 3, 5, 7, 8, 9, 10, 15, 52, 67, 79, 89, 96, 119
boardwalk, 65

boll weevil, 10, 94
bollworm, 10
Bost, P. T., 82
Bowles, David and Beth, 28
bows, 33, 36, 45
Brackettville, Texas, 5
Brazil, 21
Brazos River, 2, 52
British Broadcasting Corporation, 16
Brown, Frank, 98
Brown Schools, 59
brush management, 120
buffaloes, 4, 40, 45
Burden, W. Douglas, 61
Burleson, Edward, 54, 55, 56, 57, 58, 68, 70, 82, 86, 129
Burleson, Edward Jon, 70

calcium oxalate, 26
Caldwell County, 101, 105
caliche, 122
Callihan, Sanford, 106
camass, 33
Camp Clark, 102
*Canocanayestatetlo*, 5, 39
Canyon Lake, 9
Cape, J. M., 89
Cape's Dam, 86, 88, 89
carbon dioxide, 4
carbonic acid, 4–5
Caribbean, 21
Carriere, Solomon, 98
Catalina Island, California, 61
cattle, 10, 70, 76, 96
cattle trails, 3

caverns, 4
Chautauqua Hill, *57, 72*
chert, 33
Chihuahua, Mexico, 106
City Park, San Marcos, 76, *76–77*, 82, 89
Civilian Conservation Corps, 109
Civil War, 102, 106
Clark, Edward, 102
Clayton, J. A., 74
clearing, land, 12
Clear Springs Apartments, 71, 72
clear water flood, 7, *8*
climate, *33*, 34
Clovis projectile points, 33, 34
coastal plain, 16, 18
Cochran, Jerry, 98
Cock, Charles S., 81
Code, Tom, 69, 86
Coley, Ron, 16, 65
Colorado River, 2, 5, 40, 45, 52, 54
Comal River, 2, 3, 26
Comal Springs, 5, 118, 120
Connally, Preston and Doris, 59, 82
Cooper, Leslie, 81
Cordero y Bustamante, Manuel Antonio, 94
corn mills, 98, 100, 102, 103, 106
Corpus Christi, Texas, 98
Cortés, Hernán, 38
cotton, 10, 94, 100, 105–106
cotton gins, 3, 54, 68, 89, 94, 98, 100–103, 108, 111, *113*
cotton presses, 54, 100
cotton root rot, 10
Cottonseed Rapids, 98
Cousteau, Jacques, 16, 130
covered bridge, 113, 115, *115*
crawfish, 36
cultivation, 10, 12, 40
Cumberland Presbyterians, 103

Cummings, Ernest, 96, 126
Cummings, J.D., 96
Cummings Dam, 96, *97, 104*
cypress knees, 20

dams, 3, 7, 9, 15, 21, 23, 26, 49, 54, *54–55*, 56, 63, 67–71, *69*, 79–82, 82, 86, *86*, 89, 94, *95*, 96, 98, 100–103, *101, 104*, 105–106, *107–108*, 108, *110*, 111, *112*, 129
Dauchy, J. C., 105
Dauchy, J. M., 105
Davis, Edgar, 106, 109
de Aguayo, Marqués de San Miguel, 39, 42
de Aguirre, Pedro, 39, 40
de Alarcón, Martín, 39, 42
de Mézières, Athanase, 39, 43, *44*, 45
de Moscoso Alvarado, Luis, 39
de Rábago y Terán, Felipe, 42
de Rivera, Pedro, 39, 42
de Salinas Varona, Gregorio, 39, 40
de St. Denis, Louis Juchereau, 39, 42–43
de St. Denis, Marie Petronille Feliciane, 43
de Vaca, Cabeza, 38–39
deer, 33, 34, 36, 39–40, *42*
Delaware, 18
Del Rio, Texas, 4
detention dams, 7, 9, 82
DeWitt, Green, 49
dinosaurs, 4
dire wolves, 34
Disneyland, 63–64
diversity of life, 3, 15, *17*, 18, 26
drought, 3, 5, 24, 26, 33, 34, 56, 80, 118
Drummond, Thomas, 109
Dunn, John, 98
dwarf palmettos, 24

Earl, Richard, 9
earthquakes, 4
Edwards Aquifer, 2–5, *3*, 13, 26, 65, 79–80, 117–20, *121*, 126
Edwards Aquifer Authority, 5, 119–20
Edwards Aquifer Protection Regulations, 122, 126
Edwards Aquifer Rules, 126
Edwards Plateau, 1, 2, 4, 7, 10, 18, 33, 52, 118, *119*, 120–22, 126
Edwards-Trinity aquifer, 118
El Camino Real, 3, 39, 40, 42, 49, 67, *92*, 94
electricity, 3, 38, 54, 69, 70, 96, 98
elephant ear (taro, dasheen, or cocoyam), *20*, 26
endangered or threatened species, 5, 18, 24, 26, 27, 28, 65, 79, 117, 119–20, 122, 129
Endangered Species Act, 5, 64, 79, 119, 122
Engelke, Randy, 16, *110*
England, 7, 28, 94
Erhard, Caton, 68, 71
erosion, 4–5, 10, 12, 20, 122
Espinosa, Isidro Félix, 39, 40, *42*
estuaries, 21, 23, 120, 124
Europeans, 2, 12, 38
Evans, C. E., 72, 74, 79
Ezell, Greenberry, 69

Fairchild, Jack, 16, 122
faults, 4–5
federal acreage controls, 10
Federal Emergency Management Agency, 96, 101
Fentress, Texas, 7, 10, 12, *18, 95*, 96, *97*, 103, *105*, 105–106, 130
Fentress, James, 103
fire, 12, 36, 38, 102, 120
Firebaugh, William, 69

fish hatchery, 72, 72–74, 74, 76, 80, 129
floodplains, 9, 18, 33, 96
floods, 1–2, 7, 9–10, 18, 20, 23–24, 26, 51–53, 63, 68–69, 72, 79, 82, 94, 96, 101, 109
Florida, state of, 18, 28, 61
Florida red-bellied turtle, 28
fountain darter, 26, 28
4th Texas Infantry, 102
France and French, 39–40, 42–43, 47, 94
Franciscans, 42
freshwater inflow, 124
freshwater prawn, 20–21, 23
Frio River, 2
Fromme's ditch and garden, 57, 81

Galveston, Harrisburg, and San Antonio Railroad, 106
Galveston, Texas, 38, 94
gar, 15
Gary Field, 59, 126
George, Bill, 98
giant rams-horn snail, 28
Gipson, James, 113
Gipson's Ferry, 113
glass bottom boat, 53, 61, 62, 63, 65
Glover, Frank and Mattie, 82
Glover's Farm, 88
Glover's Island, 82
Gonzales, Texas, 10, 49, 97, 105, 109, 112–13, 115
Gonzales Warm Springs Foundation for Crippled Children, 109
Goynes, Tom, 16
grapes, 40
gravel, 12, 69, 79, 96, 98, 108, 129
grazing land, 10
Great Britain, 94
Greeks, 28

Green, Ed J. L., 70
Green, William, 70
Green Valley Farm, 96, 126
gristmill, 54, 68, 86, 94, 100, 108, 111
groundwater, 5, 10, 26, 122
Guadalupe-Blanco River Authority, 125
Guadalupe River, 3, 3, 9, 10, 21, 23, 39, 45, 49, 52, 98, 109, 113, 120, 125, 125
Gulf Coast, 18, 21
Gypsies, 57, 71
gyro tower, 4, 63

Habitat Conservation Plan, 119–20, 139–40
Hardeman, Leonidas, 101–102, 108
Hays County Courthouse, 9
Hays County Irrigation Records Book, 69–70, 96
*Hays County Times*, 7
headward erosion, 5
Hell's 1/2 Acre, 57, 81
Helotes Creek, 2
hemp, 40
hepatitis, 89, 126
highways, 12
Hill Country, 9, 13, 54, 117
Hobo Jungle, 57, 81
Hodges, Wayne, 112
Holmes, Martha Nell, 95, 101
Houston, Sam, 49, 106
hydraulic ram pump, 24, 109

ice factory, 57, 69–71
Incidental Take Permit, 119
Indiana, state of, 18
Indians: Apaches (Lipan, *apachu*), 39, 42–43, 45, 47; Aztec, 38; Bidai, 43; Cantona, 5, 39, 45; Catqueza, 45; Caynaaya, 45; Chalome, 45; Cibolo, 45; Comanche, 39, 43, 45, 47; Deadose, 42; *Diné*, 47; Ervipiame, 42; Jumano, 45; *Komántcia*, 47; Martindale projectile point, 94; Mayeye, 42; Muruam, 45; Navajo, 45, 47; Payaya, 45; Pueblo, 47; Sana, 45; Shoshone, 47; Tejas, 40, 43; Tonkawa, 43, 45; Ute, 47; Yojuane, 42, 45; Zuñi, 47
Innis, Bob, 108

Jackson, Berthold, 113
Jefferson, Sam, *81*
Jefferson, Thomas, 94
Joe's Crab Shack, 71
John J. Stokes Jr. Park, 86, *88*
Johnson, Lady Bird, 82
Jones, J. C., 96
Jones, J. J., 106
Jourdan, Z. P., 96
Julian, Isaac, 56

Katy Hole, 81
Katyville, 81
Kimball, Patsy, *95*, 105
Kyle, 5, 7

La Bahia, 42
Lamar, Mirabeau Buonaparte, 50
Landick, Steve, 98
Late Prehistoric, 34
Laws of the Indies, 47
Lea, Margaret, 106
Leffel turbines, 70, 89, 96, 100, 102, 108, *110*
Legend of Singing Wolf, 64
limestone, 1, 4–5, 7, 10, 13, 52, 62
Lindsey, William, 54, 68
lions, 34, 40
Lions Club, 7
Live Oak Springs, 52

Lockhart, Texas, 10, 94, 101, 105, 122
London, England, 7
Longley, Glenn, 118
Los Adaes, 42–43
Louisiana, state of, 18, 28, 40, 86
Louisiana Territory, 94
Lower Colorado River Authority, 96
Lowman, Al, 76, 102
Lowman, Q. J., 102
Lowman, Roston, 102
Luling, Texas, 7, 12–13, 16, 94, 96, *97, 104,* 106, *107–108,* 108–109, 111–12, 122, 130
Luling Foundation, 109

Malone, W. D., 82
Malone-Bost Dam, *57,* 82, *82, 86,* 88
mammoths, 1, 4, 33, 34, 36
Marine Studios of Marineland, Florida, 61
Martin, Joe, 105
Martindale, Nancy, *95,* 100–101
Martindale, Texas, 12–13, 16, *95,* 96, *97,* 98, 100–101, *101,* 130
Martindale Water Supply Corporation, 101
Masonic Order, 106
Massanet, Damián, 39, 40
mastodons, 1, 4, 33, 34
May, Antoinette, 111, 113
McClintock, William A., 51
McCormick wheel, 102
McCulloch's Mounted Volunteers, 71
McGehee, Albert S., 68
McGehee, Scott, 63
McGehee's Crossing, *81*
McLeod, Hugh, 49
Medina River, 2
melania snails, 28
Memorial Grove, 81
Menard, Texas, 39, 43

Merriman, Eli T., 54, 68
Merritt, Josh, 105
Merriweather, James and John, 108
Mexican War of Independence, 47
Mexico, 21, 69, 71, 106
Mexico City, Mexico, 3, 39
Miller, P. T., 72
mills, 3, 52, 54, 68–70, 79, 82, 89, 94, 96, 98, 100, 102, 106, 108–109, 111, 126, 130
Miss Hercules, 78
missions, 40, 42–43, 94
Mississippi, state of, 100
Mississippi River, 2, 52, 72
Mississippi Valley, 18
Missouri River, 2, 52
Moon, William W. and Sophronia, 54, 68
Mooney, John, 113, *115*
Mooney, Thomas, 106
Morgan Smith turbine, 70, 102–103
Moss, Randy, 118
mottes, 120
Munger gins, 103
Mynar, Fred and Brian, 98

Nacogdoches, 43
Natchitoches, 43
native species, plants and fish, 131–34
Natural Resources Conservation Service, 7
Navasota River, 39
New Braunfels, 5, 26, 52–53
New Mexico, state of, 45, 47
New Orleans, Louisiana, 2, 89
*Newton Boys,* 100
New York Zoological Society, 61
nonnative species, plants and fish, 135–37
nonpoint source pollution, 126
Nueces River, 2

nutria, 28
nutrients, 10, 13, 16, 91

obsidian, 33
oil, 12, 24, 105–106, 109, 126
oil wells, 12, 106
Old Mill Rapids, 96, 98
Old Stagecoach Road, 49
Olivares, Antonio, 40
Onion Creek, 5
opossums, 36
Orchard, John, 108
Ottine, Texas, 12, 24, 96, *97,* 109, *111,* 111–12, *112–13*
Ottine Swamp, 24, 109
Ottine Swamp Monster, 109, *111,* 112, 130
Otto, Adolf and Christine, 109, 111
overgrazing, 10, 120
oxbow lakes, 24
oxygen, 13, 21, 28

palmetto, 24, 25, 52
Palmetto State Park, 24, 25, 109, *111,* 112
Paris, France, 43
Parish, Lucious, 80
pathogens, 13
penitentiary, 56
Pennington, W. A., 56
Peppers at the Falls Restaurant, 71
Phillips, Gene, 63
plankton, 16, 24, 91
playscape, 81, *83*
plowing, 12
"plowing the river," 80
Plum Creek, 10, 105
plums, 36
*Popular Mechanics,* 61
Post Oak Savannah, 10, 12, 16
Post San Marcos, 49

pottery, 36, 38
Prairie Lea, Texas, 12, 96, *97*, 105–106
Prairie Lea Academy, 106
Prairie Lea Female Institute, 106
prickly pear, 33, 36
Purgatory Creek, 7, 9, 81

Querétaro, Mexico, 42

Rábago y Terán, Felipe, 42
Rábago y Terán, Pedro, 43
raccoons, 33, 36, 91
Rafael Rios No. 1 oil well, 106
railroads, 3, 12, 70, 81, 89, 106, 108–109, 111, 113
Ramón, Domingo, 39, 42–43
Ramon Lucio Park, 82, *88*
Reardon, Rebecca, 78
recharge, 13, 101, 118, 120, 126
recharge zone, 13, 126
Reconstruction, 106
Red Cross Life Savers, 78
Red Hill, 109
residential development, 120
Rich, Kathryn Thompson, 86–87, *87*, 89
right of capture, 5
Río de Inocentes, 39, 42
Rio Grande, 47, 94, 106
Rio Vista Dam, 7, 79–80, 82
Rio Vista Park, 15, 56, 81–82, *84–86*, 86
river activists, 7
River Corridor Ordinance, 89, 122
Riverhead Farms Inc., 70
*River of Innocence, The*, 16
Riverside Park, 57, 72, 74, , 78, 86
River Thames, 130
River Walk, 81, 89
Rockdale, Texas, 42

Roemer, Ferdinand, 52–54, 92, 113
Rogers, A. B., 56, 58–59, 61, 82, 85–86
Rogers, Paul, 56, 61, 63
Rogers Park, 57, 16, 56, 82, *84–85*
Roque de la Portilla, Felipe, 94
Ryan, Lamar, 113

saber-tooth tigers, 4, 34
Sabinal River, 2
Salado Creek, 2
San Antonio, Texas, 3, 3–5, 42–43, 45, 47, 49, 56, 63–64, 68, 71, 101, 103, 115, 117, 129
San Antonio and Aransas Pass Railroad, 109, 111, 113
San Antonio Bay, 3, 98, 120
San Antonio River, 2
San Antonio River Authority, 125
San Antonio Water System, 125
Sanborn, Chloe Walker, 72–74
San Gabriel River, 39, 42–43
San Marcos, Texas, 1, 9, 12–13, 15, 23, 26, 49, 54, 56, 59, 61, 67–68, 70–72, 76, 81–82, 85, 89, 94, 96, 97, 105, 122, 126
*San Marcos Daily Record*, 78
San Marcos de Neve, 43, *92–93*
San Marcos Electric Light Company, *69*, 70
San Marcos Gambusia, 26
San Marcos Ice Company, 70
San Marcos River Foundation, 16, 28, 75, 96, 122, 124–26
San Marcos Salamander, 26
San Marcos Sewer Company, 76
San Marcos Springs, 3–5, *6*, 9, 34, 36, 38–40, 43, 47, 49, *50*, 58, 66, 118–20
San Marcos Utilities Company, 56, 74

San Marcos Water Company, 70
San Saba River, 39, 43
Sansom, Andrew, 66
Santa Cruz de San Saba mission, 43
Sargasso Sea, 21
sawmills, 52, 54, 68, 86, 89, 94, 108, 111
Seadrift, Texas, 98
sea levels, 4–5
sea lions, 61–63
sediment, 9–10, 13, 16, 91–92, 96, 129
septic tank systems, 13
Sessom Creek, 71, 129
Sewell, S. M., 72, 74, 75, 78
Sewell Park, 75
Shiner, Joel, 34
Short, Brewster, 112
Sibley, H. H., 106
Sierra Club, 5
Silver Springs, Florida, 61
Sink Creek, 7, 9, *50*
sinkholes, 5, 7, 13
skating rink, 105
sky ride, 63, 89
slaves, 47, 86, 89, 94, 106
Slayden, James Luther, 113
Slayden, Texas, *97*, 113, *114*
Smith, A. H., 100
Smith, C. Spurgeon, 72
Smith, Cullen, 103
Smith, Joseph, 103
Smith, W. R., 105
Smith, W. S., 98
snakes, 15, 24, 36, 91
sorghum mills, 100
Southwest Texas State Normal School, 72
Southwest Texas State Teachers College, 67, 74
Spanish, 1, 3, 39–40, 42–43, 45, 47, 49, 52, 69, 94, 105

*Index*

Spanish missionaries, 42
Spring Lake, 15, *21*, 26, 28, 43, *50*, *55*, *57*, *58*, 59, 61–62, 64, 69
Spring Lake Hills Garden Club, 81
Springlake Marine Gardens, 61
Spring Lake Park Hotel, *57*, *58*, *59*, 86
squirrels, 36
Stair, Oscar Fritz, 106
Stairtown, Texas, 106, 109
Staples, John Douglas, 102
Staples, Texas, 12, 96, *97*, 101–103, *102–104*, 105, 130
Staples Water Power Company, 102
storm water runoff, 13
Strahan Coliseum, 76
stream piracy, 4, 5, 89
sugarcane mills, 102
Sunset Branch, Southern Pacific Railroad, 108
Supple, Jerry, 65
suppressing natural fires, 12, 120
swamp rabbits, 20, 23
swimming pigs, Magnolia and Ralph, 63–64, 89
Syers, Ed, 112
synthetic fibers, 10, 94

Tarzan, 63
Taylor, T. U., 70, 79, 89, 98, 100, 102–103, 106, 108, 111
Tegrotenhuis, Duane, 16
Teller, J. W., 100
Terán de los Ríos, Domingo, 5, 39, 40, 45
Texana Village, 63
Texas Agricultural Experiment Station, 26
Texas A&M University, 16
Texas Blind Salamander, 26, 65
*Texas Chainsaw Massacre, The*, 100

Texas Commission on Environmental Quality, 125–26, 130
Texas Legislature, 5, 69, 125–26
Texas Natural Resource Conservation Commission, 125
Texas Parks and Wildlife Department, 18, 66, 89, 118, 122, 124, 130
Texas Power and Light Company, 96
Texas Rivers Center, 6, *59*, *60*, 66, 89
Texas State Demographer, 101
Texas State University–San Marcos, 9, 16, 18, 43, 64–65, 72, 89
Texas Supreme Court, 122
Texas Water Conservation Association, 125
Texas Water Development Board, 125
Texas Water Safari, 16, 98, *99*
Texas Water Trust, 125
Texas wildrice, 26, 27, 79–80
32nd Texas Cavalry, 102
Thomas, Denny, 122
Thompson, William A., 86, 89
Thompson, William Hardeman, 89, 126
Thompson's Islands, 67, 70, 86, *87*, 89, 94, 96
time periods: 33; Paleo-Indian, 33, 34, *35*; Early Archaic, 33, 34, *35*, 94; Middle Archaic, 33, 34; Late Archaic, 33, 34; Late Prehistoric, 33, 34
Tips, Walter, 81
Tolhurst, C. E., 105
tortoises, 33, 36
tourism, 65, 69, 89
toxic materials, 13
transportation, 2–3, 52, 67–68, 106
trees: American elm, 18; baldcypress, 18, 20, 22, *32*; black willow, 18; bois d'arc, 89, 96; box elder, 18;

Carolina ash, 18; cedar (juniper), 10, 69, 74, 76, 94, 96; cedar elm, 18; cottonwood, 18, 69; mulberry, 40; oaks (various), 10, 18, 33, 34, 36, 51, 120; pecan, 12, 18, 33, 34, 89; sycamore, 18; walnut, 18, 69, 89, 111–12; willow, 74
Trinity River, 39
turbines, 3, 69–70, 89, 96, 98, 100, 103, 106, 108, 111
turtles, 15, *20*, 28, 33, 36
Twain, Mark, 117

underwater theater, 61
University of Texas–Austin, 18, 70
Upper San Marcos Watershed Reclamation and Flood Control District, 7
U.S. Army Corps of Engineers, 96, 130
U.S. Department of Commerce, Bureau of Fisheries, 74
U.S. Environmental Protection Agency, 130
U.S. Fish and Wildlife Service, 5, 18, 24, 119–20, 129–30
U.S. Geological Survey, 7, 70
Ussery, Ab, 112
U.S. Soil Conservation Service, 7
U.S. Supreme Court, 119–20
Uvalde County, 117

Vail of Avoca, 89
Veramendi, Juan, 68
Victoria, Texas, 3
Vogelsang's Lodges, *57*, 71, 72
volcano, 63
Von Roll Iron Works, 63

Waco, Texas, 3–4
Walker, J. K., 108

Walling, George, 62
Ward, Robin, 64
Warner, Jack, 61
Wassenich, Dianne and Tom, 16
wastewater, 13, 59, 76, 96, 101, 122, 124, 126; activated sludge treatment plant, 57, 76
water hyacinth, 28
Watermelon Thump festival, 106
water power, 70, 102
*Water Powers of Texas, The,* 70, 98
watersheds, 1–2, 7, 10, 12–13, 52, 126

watertrumpet, 28
Webb, Billy, 112
Weeki Wachee Springs, Florida, 61
Weissmuller, Johnny, 63
Westerfield Crossing, *93*
Westinghouse, 70, 98, 100
Willow Creek, 7
Wimberley, C. W., 67, 69, 70–71, 79, 81–82
Wimberley, Pleasant, 67
Wimberley, Rufus, 69, 74
Winn, Buck, 63

Wonder Cave, 56
Wood, Peter C., 102
Woodruff, Charles, 5
World War I, 111
World II, 105, 113

York Creek, 7, 10

Zedler, Berthold, 111, *112–13*
Zedler, Fritz, 108, 111
Zedler's dam and mill, 16, *104, 107–108, 110*

ISBN-13: 978-1-58544-542-4
ISBN-10: 1-58544-542-8